Second Nature
Economic Origins of Human Evolution

Was exchange an early agent of human evolution or is it merely an artifact of modern civilization? Spanning 2 million years of natural history, this book explores the impact of economics on human evolution. The theory of evolution by natural selection has always relied in part on progress in areas of science outside biology. By applying economic principles at the borderlines of biology, Haim Ofek shows how some of the outstanding issues in human evolution, such as the increase in human brain size and the expansion of the environmental niche humans occupied, can be answered. He identifies distinct economic forces at work, beginning with the transition from the feed-as you-go strategy of primates through hunting-gathering and the domestication of fire to the development of agriculture. This highly readable book will inform and intrigue general readers and those in fields such as evolutionary biology and psychology, economics, and anthropology.

HAIM OFEK is Professor of Economics at Binghamton University, NY.

Second Nature

Economic Origins of Human Evolution

Haim Ofek

Department of Economics
Binghamton University

CAMBRIDGE
UNIVERSITY PRESS

PUBLISHED BY THE PRESS SYNDICATE OF THE UNIVERSITY OF CAMBRIDGE
The Pitt Building, Trumpington Street, Cambridge, United Kingdom

CAMBRIDGE UNIVERSITY PRESS
The Edinburgh Building, Cambridge CB2 2RU, UK
40 West 20th Street, New York, NY 10011–4211, USA
10 Stamford Road, Oakleigh, VIC 3166, Australia
Ruiz de Alarcón 13, 28014 Madrid, Spain
Dock House, The Waterfront, Cape Town 8001, South Africa

http://www.cambridge.org

First published 2001

Printed in the United Kingdom at the University Press, Cambridge

Typeface Swift (Adobe) 9.5/14pt.   System QuarkXpress® [wv]

A catalogue record for this book is available from the British Library

Library of Congress Cataloguing-in-Publication data
Ofek, Haim, 1936–
    Second Nature: economic origins of human evolution/Haim Ofek.
      p. cm.
    ISBN 0 521 62399 5 – ISBN 0 521 62534 3 (pbk)
    1. Human evolution. 2. Economics, Prehistoric. 3. Commerce, Prehistoric.
    4. Economic history. I. Title.
    GN 281.4.O35 2000
    306.3′093–dc21   00-036293

ISBN 0 521 62399 5 hardback
ISBN 0 521 62534 3 paperback

Contents

Acknowledgments

Figure 2.1 (p. 16): Courtesy of Priscilla Barrett.

Figure 4.2 (p. 48): Copyright 1978 by Princeton University Press. Reprinted by permission of Princeton University Press.

Figure 5.1 (p. 67): Copyright 1995 by The Wenner-Gren Foundation for Anthropological Research. Reprinted by permission of the University of Chicago Press.

Figure 5.2 (p. 81): Copyright 1965 by Princeton University Press (renewed 1993). Reprinted by permission of Princeton University Press.

Figure 7.1 (p. 112): Copyright 1995 by Princeton University Press. Reprinted by permission of Princeton University Press.

Figure 11.1 (p. 171): Panels (a) and (b) adapted from *Field Guide to Early Man* by David Lambert and the Diagram Group. Copyright 1987 by Diagram Visual Information Ltd. Reprinted by permission of Facts on File, Inc. Panel (c) Copyright 1998 by the Board of Trustees of the University of Illinois. Used with the permission of the University of Illinois Press.

Figure 11.2 (p. 175): Sea Level insert with permission from D. Raynaud *et al.* (1993, Figure 5). Copyright 1993, American Association for the Advancement of Science.

Figure 12.1 (p. 197): Copyright 1993, American Association for the Advancement of Science.

Figure 12.2 (p. 204): Used with the copyright permission of *Nature*.

1   Introduction

The propensity and capacity to exchange one thing for another between two traders – however unrelated to each other – is a profound distinguishing feature of human subsistence. Human beings are endowed with remarkable skills of trade which they deploy spontaneously when confronted with favorable opportunities; skills that lie dormant in the absence of such opportunities. As is true of other innate human abilities – such as the mastery of spoken language – basic skills of trade are taken for granted precisely because they are either inborn or acquired at a young age without conscious effort. Such skills are not as trivial as they may seem to a casual observer or, for that matter, to their very practitioners. Exchange requires certain levels of dexterity in communication, quantification, abstraction, and orientation in time and space – all of which depend (i.e., put selection pressure) on the lingual, mathematical, and even artistic faculties of the human mind. Moreover, exchange relies on mutual trust: predictable codes of conduct agreeable to the human sense of morality. Exchange, therefore, is a pervasive human predisposition with obvious evolutionary implications. The root cause of this predisposition and its evolutionary consequences in history, and prehistory, are the central concerns of this book.

Was exchange an early agent of human evolution, or is it a mere *de novo* artifact of modern civilization? The evolutionary literature treats the question with great caution. Many authors, starting with Charles Darwin and Alfred Russel Wallace, preferred to avoid the issue altogether. When the issue comes to the fore, the importance of exchange in recent industrialized societies is readily acknowledged. However, its importance in any but the most recent stages of human history is typically dismissed. In its present status, human exchange is in the same state of scholarly inquiry as human language was just a century ago (when conventional wisdom recognized sophisticated linguistic forms only in modern civilizations). Conventional wisdom today seems to suggest that human exchange is essentially an incidental by-product of previously evolved mental and social (or even cultural) structures, rather than a distinct agent of evolution. The discussion throughout this book calls into question the merits of this article of conventional wisdom in view of, among other things, *Darwin's principle of utility* and *Wallace's independent proof*, two sources of difficulty in the study of human evolution from its very dawn in classical Darwinism.

Both Darwin and Wallace were keenly aware of certain structures and refinements of human intelligence which are (seemingly) unaccounted for by natural selection. Each in his own way made equally unsuccessful attempts to identify the missing agent. Wallace's attempt though, was bolder and in the end more embarrassing. The main difficulty was presented by what seems to be a premature and excessive advance in cognitive skills relative to prehistoric needs for human survival. What useful function could the higher faculties of the human mind (like mathematics and music) serve at the stages in human evolution in which they evolved? No good explanation compatible with the demands of natural selection was available either to Darwin or to Wallace, and none has yet been offered. Yet, all the while a plausible explanation was brewing within reach. The full account sounds much like a story of a missed opportunity (to be told in Chapter 3).

From its very inception, the theory of evolution by natural selection has been tormented by frustrating puzzles, not the least, the one just outlined. Many of these are clearly ascribable not so much to lack of evidence as to the availability of evidence that defies interpretation. With the great benefit of hindsight, it is now also clear that the failure in interpretation itself was on many occasions (but not always) due to lack of progress in adjacent fields of science. The age of the earth and the geographic distribution of species were two fiercely challenging puzzles that baffled Darwin to his last days. Both have since been fully resolved in his favor, albeit decades later – the former with the discovery of radioactivity and the latter with the discovery of plate-tectonics. Darwin's triumph (in bequest) was thus accorded, in these two particular instances, not so much by new evidence from within the field of evolution as by belated progress from without – in the fields of physics and geology, respectively. At issue in this volume are outstanding questions in human evolutionary history and the attempt to resolve them with the aid of insights from yet another field, perhaps not closely adjacent to evolution, but at least tangential to it: economics.

Some of the outstanding issues, and puzzles, in the field of human evolution possess a deep economic dimension that is not always fully recognized as such. Examples range from the most general issue associated with the evolution of the human intellect mentioned above to more narrowly focused issues that are equally puzzling, and equally unresolved. Consider some unexplained remarkable facts:

- An allegation of premature development is held not only against the higher faculties of the human mind, but also against the human faculties of making fire. Even by the most conservative estimates going back only 300,000 to 400,000 years (others put it at 1.5 million years and more) the deliberate use of fire by humans represents a considerable technological advance over stone tool manufacturing or, arguably, even over the invention of the wheel (dated, by comparison, only 5,000 years ago). In other words, domestication of fire seems to enter the record unexpectedly ahead of its time.

- Caches of finished stone tools as well as raw material from distant sources of flakeable rocks (10 kilometers or more away) were found in several early hominid East African sites dated between 1.5 and 2 million years ago. Could a hominid with a brain half the size of a modern human have the resources (and foresight) to maintain an inventory of raw materials? If so, what could possibly be the principle of economic organization under which such a practice was motivated, and such redundancy afforded?

- The human gut is markedly small relative to body size and in proportion to similar metabolically expensive organs in the human body: the heart, liver, kidneys, and lungs – not to mention the brain. In fact, it has been estimated (Aiello and Wheeler, 1995) that the total mass of the human gastrointestinal tract is only about 60% of that expected for a similar sized primate. By these standards, human gut dimensions are those of a meat-eater (Chivers, 1992). Yet, world wide, meat usually constitutes only a small proportion of the total human intake of food. This raises a serious question: the compatibility of an organ with its primary function.

- The Upper Paleolithic people (roughly, 40,000–10,000 years ago) greatly extended the geographic distribution of humankind to include easternmost Europe, northern Asia (Siberia), Japan, Australia, and the Americas. But the major thrust was largely inland rather than overseas with eastbound migration flowing from central Europe toward Asia and northbound migration moving on both continents toward the arctics. Were these people heading in the wrong direction in the midst of an ice age?

- The *Iliad*, the first known masterpiece of western art (literary or otherwise), is a war story. Warfare in all its glory and horrors has been repeatedly depicted (and indicted) in future generations as well: *Henry V, Wellington's Victory, War and Peace, Battleship Potemkin, Guernica* – are but a few reminders that this theme is part and parcel of civilized artistic expression as much as warfare and interpersonal violence are part and parcel of civilization itself. Against all preconceptions, the theme is almost invariably absent from all expressions of prehistoric art. Cave paintings and contemporaneous portable art rarely show men or, for that matter, women in combat. Nor does the corresponding fossil record show much in the way of numerous broken human bones or any other compelling skeletal evidence for deliberate injury (these start to appear with any regularity only with agriculture). Is it safe to assume that these early hunter-gatherers "could not afford the kind of risk-for-limited-return involved in hunting their neighbors" (Klein, 1989)?

- No species has ever been observed to abruptly desert the niche it occupied in the environment in favor of another. Yet this is precisely what transpired in the great human transition to agriculture that took place almost simultaneously in widely separated parts of the world, for no apparent reason. Of these, the dual origin of agriculture (in the Old and New World) is the most puzzling of all.

- Husbandry is a labor-intensive undertaking. It takes in general more time and human effort to raise and slaughter a domesticated animal than to hunt and kill its wild counterpart. One lucky strike with an arrow can earn an expert hunter the same amount of meat and nearly all the byproducts (skin and fiber, though not milk) that a herder will obtain only by long hours of toil over months if not years in waiting. It is thus difficult to understand how pastoralism could have so completely displaced hunting to begin with. Why did humans for the first time, and of all times, choose to rely on domesticated stock precisely when (due to a climatic optimum) wild stock in many parts of the world was more abundant than ever?

To be sure, outstanding issues like these come with their own peculiarities and, as such, are treated in the pages of this volume on a case by case basis: with evidence (when available), with logical inference (when applicable), and – only as a last resort – with conjecture. But they also share a common core that calls for a unified treatment and, perhaps, a unified explanation.

The difficulty in reaching a unified explanation can be traced in part to the relative neglect of economic reasoning in the way we tend, all too often, to approach the affairs of our own ancestors – however remote. Economic principles are not designed for the sole use of modern people. In the application of economic principles or, for that matter, evolutionary principles to the affairs of early humans it is useful to recognize two sweeping trends in their (and our own) evolution: the expansion of the brain and the expansion of the niche. The persistent expansion in brain size is by far the most impressive evolving anatomical trend that, by the very nature of the organ, far exceeds anatomy itself. The ever-expanding niche that humans occupy is the most impressive evolving trend from the viewpoint of economics, for economics is fundamentally the study of niche expansion. The remaining challenge is to make the necessary connection.

Part 1
Bioeconomics

2 Exchange in human and nonhuman societies

Upton Sinclair's novel, *The Jungle* (1906), is a brutally graphic account of the ruthless competition in the stockyards and slaughterhouses of Chicago at the turn of the nineteenth century. Literary observers like Sinclair, and social observers in general, have often appealed to an imaginary animal-like struggle for survival in search of analogies that describe human conduct in the marketplace. The analogy is unfair to humans as much as to animals. In reality, the essential pattern of market activities, perhaps more than any other pattern of human behavior, is marked by the *lack* of analogy with animals.

Exchange, or apparent exchange, among living organisms other than humans is largely confined to the realms of *symbiosis* and *nepotism* (i.e., transfers among members of separate species and transfers among *related* conspecifics, respectively). For human beings these two patterns of exchange are only part of a wider repertoire that includes a remarkable addition in the form of *mercantile exchange* (transfers among conspecifics at large). A preliminary survey of these three patterns of exchange will be given in this chapter.

Adam Smith's zoological digression

Adam Smith was a younger contemporary and, it is told, a great admirer of Linnaeus (Schabas, 1994:332). The Linnaean version of the "economy of nature" had already acquired some enthusiastic following among English-speaking readers like Erasmus Darwin, another contemporary of Linnaeus (and grandfather of Charles), who cast the Linnaean system into verse under the title *The Botanic Garden* (1789). Smith's main concern, however, was the man-made "political economy." It was natural for him to point out a fundamental distinction (one of many) between the two systems:

> Nobody ever saw a dog make a fair and deliberate exchange of one bone for another with another dog. Nobody ever saw one animal

by its gestures and cries signify to another, this is mine, that
yours; I am willing to give this for that. (1976:17)

This remark was meant to emphasize – by *lack* of analogy – the unique
manner in which exchange operates in human affairs. It denies neither
the existence nor the prevalence of exchange elsewhere in nature.
"When an animal wants to obtain something either of a man or of another animal it has no other means of persuasion but to gain the favour of
those whose service it requires," he states and adds the pivotal insight:
"Man sometimes uses the same arts with his brethren" (1976:18). In other
words, Adam Smith suggests two distinct mechanisms of exchange. First,
a fairly formal mechanism exclusive to humans that operates "by treaty,
by barter, and by purchase" (1976:19). The second is a universal mechanism common to humans and animals, relying – as Adam Smith saw
things – on benevolence induced by begging, essentially, on emotional
currency.

Adam Smith deserves some credit for noticing a pattern of begging in
animal exchange. Manifestations of infantile and submissive modes of
behavior (typical of mammals and birds) are prevalent between the young
and their parents, between mating partners, and among members of
packs, flocks, and other group formations in which members react to one
another on the basis of individual recognition. The most obvious examples among mammals include some free-living relatives of the domesticated dog (wolves, foxes, bush dogs, and above all, African wild dogs) and
to a lesser extent man's own relatives (the great apes and other primates).
It does not take long to recognize the interplay of these preadaptations of
begging and submissive behavior in the relationship between dogs and
man, a relationship Adam Smith used as an illustration.

The main difficulty with Adam Smith's account of animal exchange,
however, is that it relies on sentiments. Counter examples are easy to
come by. Modern observers of animal behavior may call attention to
instances in which exchange operates flawlessly by rigid stimulus-
response mechanisms, or by outright reflex, rather than by cognition
and sympathy. Little or no begging or benevolence is evident in the
exchange between a bumblebee and the plant it pollinates or, for that
matter, in exchanges among workers in a colony of social insects.
Nestmates in a colony of ants, for instance, typically exchange liquid

food through regurgitation induced by a recipient touching her forelegs to the donor's head. A casual observer may view the event as an encounter between sisters that have the capacity to express and compassionately attend to each other's needs. However, as Hölldobler and Wilson report, the processes can be simulated mechanically by touching the same spot on the donor's head with a fine human hair. The ant will respond by regurgitating in front of its human handler (Hölldobler and Wilson, 1994:51). Evidently, a regurgitating ant scarcely exceeds the level of compassion expected of a vending machine. Exchange is effected, in this instance, by some sort of a vomit reflex rather than by emotional currency.

Adam Smith's skills as an observer of animal affairs apparently fell short of his skills as an observer of human affairs, but his mistakes should not be taken as an excuse to ignore his larger issue. In the end, his main assertions (starting with the one quoted above) echo fundamental evolutionary dilemmas that should challenge observers of humans and observers of animals alike. At issue are the peculiarities and origins of human exchange and, by implication, of division of labor and human diversity. To understand these peculiarities and fully appreciate his evolutionary dilemmas, one needs to refine and sharpen the borderline Smith himself sought to draw between animal exchange and human exchange.

Symbiotic exchange

Symbiosis is the phenomenon of reciprocal and mutually beneficial transfer – or "exchange" – of resources and services across species.[1] Some economists view it as the closest thing to formal trade that involves nonhuman players (e.g., Tullock, 1994:83). Agriculture and, in general, the relationship between humans and domesticated plants and animals is a primary example of *symbiosis*. Obviously, it is hardly an exclusive example.

The most commonly observed examples of symbiosis among nonhuman parties are animal–plant relationships (e.g., between the fig and the

---

[1] The term *symbiosis* is used here in a restrictive sense. It includes only interspecific relationships that benefit both sides. *Parasitism* and other forms of cohabitation that benefit only one partner are excluded.

fig wasp). Animals and plants best meet the two salient prerequisites of symbiosis: the parties do not compete for the same resources and they tend to make up for each other's shortcomings. Plants typically provide food and shelter in return for pollination, dispersal, fertilization, pest control, and so on. Symbiosis on a grander order of ecological organization takes place, of course, between all animals *as a group* and all (green) plants *as a group* through the exchange of oxygen for carbon dioxide in the atmosphere at large.

Symbiotic relationships between animals are less abundant. A conspicuous example in this category is cleaning symbiosis. Most prevalent among fish and other forms of marine life, cleaning is occasionally observed among land animals as well (e.g., between birds and ungulate animals).[2] Relationships between macro- and microorganisms (e.g., between termites and cellulose-digesting microbial symbionts that live in their guts) are probably more abundant, though obviously less conspicuous. But the most consequential of all the symbiotic relationships is reserved to exchange among microorganisms themselves. It may occur when one cell is engulfed by another, but instead of being digested by its host, the two initially establish a stable relationship of intracellular symbiosis and eventually become fused. In the end, this process of evolution (by infection) results in a new more complex cell. By establishing this process of symbiogenesis, Lynn Margulis (1981) managed almost single-handedly to resolve one of the greatest mysteries in the history of organic evolution: the breakthrough emergence of cells equipped with organelles. On the whole, the list of ecological phenomena under the heading of symbiotic exchange is open ended, assuming one is willing to use the term *exchange* in a figurative sense.

Nobody can hold Adam Smith liable for all the figurative deflections of the word "exchange." Economists are reluctant to use the term in any but the strict sense (that is, when the transaction is made voluntarily and deliberately by the parties), and symbiotic relationships hardly apply. From an economic point of view symbiosis is little more than a procedure for acquiring resources from the environment, like grazing or, for that matter, mining. It is true that close proximity between "host"

---

[2] Not counting oxpeckers. Recent work suggests that oxpeckers get a large part of their daily food intake from blood, keeping old wounds in their hosts' skin open, or indeed inflicting new wounds (Weeks, 1999).

and "client" occasionally calls for refined skills of recognition and eva-
sion in case one of the parties is a disguised predator or noxious creature
– on this account, the analogy with human exchange is not so easily dismissed. Thus, individual recognition and even individualized pairing is
occasionally evident (e.g., in cleaning interactions), apparently, in order
to safeguard the approach and close contact between the parties. The
essential point, however, is that symbiosis is a racial rather than individual experience. Though all transfers in typical symbiotic exchanges
occur between individual organisms, the volume and attributes of
resources exchanged are collectively regulated, in the evolutionary sense
(and time scale) at the level of populations and species or even higher up
in the ladder of organization, rather than at the level of individual
organisms – let alone individual transactions. As such, symbiosis leaves
room neither for competitive bargaining nor for free interplay of strategic behavior between traders. Exchange in the sense of physical transfer
(delivery and collection, reciprocal or otherwise) is exposed to natural
selection, but exchange in the strategic sense is largely shielded from it.
Little selection pressure is consequently exerted at the level of intelligence of the parties beyond the call of one-sided smart procurement, as
distinct from trade which is a two-sided activity. Symbiosis has apparently equipped its participants with the brains of resourceful harvesters
rather than with the brains of shrewd traders. Above all, symbiosis is typically confined to the transfer of a particular resource or service in a well-
defined environmental setting. In contrast, exchange (as economists
make sense of the term) needs to entertain from time to time novel commodities in new settings.

Though symbiosis lacks a strategic dimension, it cannot be completely dismissed as an economically meaningless form of exchange. At least
on the grounds of division of labor, a highly relevant facet of Adam
Smith's analysis, symbiotic exchange has a fairly compelling economic
meaning. The primary function of exchange, Adam Smith emphasized,
is to promote the division of labor (mutually beneficial differentiation
and specialization of function and form). Symbiotic exchange promotes
division of labor between species. It does not promote, however, division
of labor among individuals within a species. The task of promoting
intraspecific division of labor is left to exchange among members of the
same species.

Kin and nepotistic exchange

Going beyond symbiosis, what are the advantages of exchange within a species, and to whom do they matter? Domesticated animals have little use for such advantages because humans control the distribution of resources among them, from food to shelter to territory, and humans also coordinate their vital activities: choice of mates, parental efforts, rest, herding, and defense against predators – to mention only some functions of exchange. Thus, domesticated animals have long been shielded from selection pressure promoting exchange among conspecifics (a fact that helps to explain the reduced brain volume in such animals compared with their relatives and progenitors in the wild). Adam Smith's sweeping assertion that "Nobody ever saw a dog make a fair and deliberate exchange of one bone for another with another dog" should come as no surprise: the dog is one of the oldest species under domestication.

Exchange among conspecifics is not so easily dismissed in the wild. To free-living animals the advantages of exchange are often too great to be ignored. The most obvious of these advantages are associated with one or more of the following three functions:

- *Redistribution* results in more efficient utilization of acquired or otherwise available resources. Many animals – notably carnivores and espe-

---

Box 2.1 The vampire bat: subsistence verging on trauma

Equipped with razor-sharp teeth, the vampire bat (*Desmodus rotundus* among others) feeds on blood of cattle and other livestock. As a staple food, blood is rich in protein but short on calories – the wrong cocktail for a high-energy flying animal. Consequently, the vampire bat must consume more than half its body weight in blood daily. It cannot survive more than 48 hours without a drink. A female bat that could not find a victim on any given flight can usually obtain its life-saving meal of regurgitated blood from a more fortunate female. On another day, the roles may alternate. For further discussion of this exceptional case, where subsistence approaches the brink of trauma, see Box 2.3.

cially scavengers – rely on food-sharing for subsistence. Humans, the most carnivorous primates, are not exempt. Archeological findings (e.g., Isaac, 1989) suggest that the ritual of food-sharing was probably practiced by premodern humans on a regular basis. More recently, direct observations by anthropologists (especially, Hill and Kaplan, 1994) further suggest that food-sharing is not merely an occasional ritual but a central organizing principle in human affairs. The function of food-sharing in tribal societies that rely on hunting for subsistence, bears a close resemblance to the function of financial markets in societies that rely on formal trade. The importance of food-sharing to certain species other than humans may even transcend the normal boundaries of subsistence. Thus, for the vampire bat redistribution through food-sharing is virtually a life-preserving device (see Box 2.1).

• *Division of labor* facilitates efficiency in procurement and processing of resources through differentiation and specialization of function and form. Division of labor between the sexes, or among castes, in tasks associated with parenting, guarding, food acquisition, food transport, and nesting is fairly common in organisms that care for their young. Group-living animals may further rely on division of labor in more subtle tasks associated with various roles of leadership, subordination, risk taking, and risk shifting. A straightforward example of task (and team) oriented division of labor is provided by the "digging chain" performed by the naked mole-rats (*Heterocephalus glaber*) in their mining activities (Figure 2.1).

• *Cooperation* widely opens the way to synergy: amplification of individual efforts through collective action. It promotes, for instance, the formation of teams of small predators to attack large prey and, conversely, formation of teams of small prey to fend off large predators. In both cases, cooperation affords the advantage of maintaining economical body size.

Some of these advantages initially occur, of course, on the level of groups and populations rather than on the level of individuals. However, in each of the functions listed above, exchange is either an integral part of the activity or its prerequisite. Since exchange is a mechanism by which advantages gained at the level of the group may be internalized by its individual members, it follows that a plausible selection pressure may every now and then, and in varying degrees, establish (in an evolutionary sense) adaptive patterns of behavior compatible with redistribution, division of labor, and cooperation. Such patterns are most apparent in the behavior of social insects among the invertebrates, and in the behavior of certain species of mammals (including humans) among the vertebrates.

Among nonhuman mammals, there are certain solitary species (notably, the naked mole-rat mentioned above) and at least one wider group of related species (in the *Canidae* family) that routinely rely on all three functions outlined above. The *Canidae* family includes the wild progenitor of the domesticated dog, presumably the gray wolf, and some of its close relatives (foxes, jackals, bush dogs, and the like). Patterns of exchange designed to take advantage of efficient redistribution, division of labor, transport, and cooperation are typical of many species in this

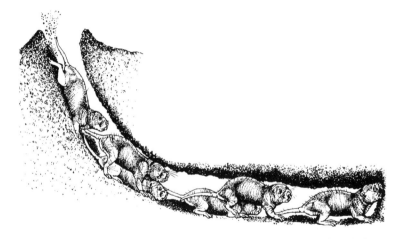

Figure 2.1 **Division of labor in mole-rat mining** A digging chain typical of the naked mole-rat (*Heterocephalus glaber*). A number of naked mole-rats cooperate in digging, performing different tasks. The animal in front is responsible for excavation using its protruding chisel-like incisors for digging. Once a pile of soil has accumulated, it brings both its hind feet forward, collects the soil, and kicks it backwards. The animal first in line behind the digger receives the soil and pulls it behind itself reversing along the burrow, the body close to the floor, until it can pass the soil to the animal responsible for soil dispersal at the tunnel's entrance. It then returns to the front, straddling the line of mole-rats pushing soil backwards, to start a new round. It should be noted that mole-rats easily and rapidly travel backwards in their compact burrows, where turning is often impossible (see also Box 6.1). From Macdonald 1984.

Box 2.2 The wild African dog

The wild African dog (*Lycaon pictus*) is probably the most highly social species within the *Canidae* family, and in many ways the very model of a sociable mammal. Members of this species live in tight packs of closely related males (and migrant females) headed by a single breeding pair. They rear their young and hunt cooperatively, and are generally engrossed in a lifestyle that entails intensive food-transport and food-sharing. A dog in dire need of a meal can acquire food from other members of the pack by means of infantile begging or display of active submission that seems to persuade fellow dogs to share kills or even disgorge recently eaten meat. The roles of donor and recipient alternate frequently among able adults. Such a pattern of food-sharing seems to suggest exchange of status (hierarchical rank in the pack along with prospects of breeding) for food.

Division of labor in energetically costly roles such as parenting, den guarding, and hunting, in combination with the ability to travel at high speeds for prolonged periods of time, enables the pack to operate over vastly extended food-sparse areas (typical home ranges are 450–2000 km^2). Special adaptations for food retrieval and transport further enable them to collect and return available food items to the lactating female and her altricial offspring which, at all times, are sequestered and guarded at fixed locations of maximum safety. Cooperative hunting, in turn, allows these lightweight carnivores (17–36 kg) to extend their sources of food to prey as heavy as zebra (200 kg). Food-sharing enables efficient utilization of large carcasses and the distribution among pack members is generally in tune with the rules of optimal investment. For instance, yearlings (that already represent a large investment of the pack) are given priority in feeding over pups (a smaller past investment). Similarly, the fact that pack members share food and provide care for sick adults represents an attempt to protect and, occasionally, recover even larger past investments (Wilson, 1975; Moehlman, 1989; and Sheldon, 1992).

family, but are probably most apparent in the behavior of the wild African dog (see Box 2.2). Food-sharing among pack members of these wild dogs bears a close resemblance to a form of exchange, not only because the roles of donors and recipients frequently alternate, but primarily because it facilitates an efficient system of transportation and redistribution of perishable resources.

For all their apparent versatility, the privilege of exchange among wild dogs is extended however only to fellow pack members: to close relatives and to breeding individuals who may produce or tend closely related pups. No strangers need apply. The same nepotistic pattern of exchange repeats itself (usually to a lesser degree) in other mammalian and avian species, and (to a much higher degree) in insects and other invertebrates.

While humans associate exchange with subsistence or accumulation of wealth, the key to animal exchange (among conspecifics) is procreation and kinship. Unlike human transactions that are often (but not always!) determined by the merits of the entities exchanged, transactions among animals are almost invariably determined by the identity of the trading parties or by the gregarious function of the interaction between them. Sometimes the entity exchanged completely loses its meaning. The transaction becomes a ritual. For instance, courtship feeding is a form of exchange widespread among mating birds and often no actual transfers of food take place (Immelmann and Beer, 1992:63).[3] As with wild dogs, intraspecific transfers in other nonhuman societies take place only among select members (closely related individuals or mating partners) or in the context of well-defined functions confined to the realms of procreation or kinship. Examples of voluntary transfers outside these realms are rare (see Box 2.3 for some exceptions that prove the rule). It is thus appropriate to qualify this pattern of transactions by the term *kin* or *nepotistic exchange*.

In one form or another, and to different degrees, nepotistic exchange (such as mutual aid among family members) is an integral part of the life strategy for many organisms. Humans are not exempt: In isolated tribal societies it is practically the only way by which exchanges are affected, simply because traders are recognized by kinship. The subtle pattern of

---

[3] The logic of this ritual is not confined to animal exchange. Reports of people exchanging like items (pigs for pigs) can be found in the anthropological literature, for instance, in reference to the ceremonial exchange system in Melanesia (Strathern, 1992:170).

Box 2.3 Business is business

Altruism in the realms of subsistence is exceedingly rare in nature. In every case known to me it seems possible to claim that voluntary transfers (to unrelated individuals) are associated with life-threatening contingencies, distinct from ordinary routines of subsistence. This includes the case of the vampire bat discussed above (Box 2.1) and most of the examples cited in the literature dealing with reciprocal altruism (e.g., Trivers, 1971, 1985). An instructive example is provided by species of penguins and other birds that practice communal defense of their breeding colonies. These birds may risk their lives fending off predators in an attempt to save temporarily abandoned or orphaned unrelated young, but will remain "indifferent" to the plight of the same chicks when they starve to death (Immelmann and Beer, 1992:64). As Darwin once relayed, a similar pattern can be seen amongst monkeys in captivity:

> Orphan monkeys were always adopted and carefully guarded by the other monkeys, both males and females. One female baboon had so capacious a heart that she not only adopted young monkeys of other species, but stole young dogs and cats, which she continually carried about. Her kindness, however, did not go so far as to share her food with her adopted offspring. (Darwin, 1874: 70-71)

---

gift-giving described by Marcel Mauss is a typical mechanism of exchange in such societies. According to Mauss (1967) the exchange of goods through gift-giving is not merely a mechanical institution but a moral transaction, bringing about and maintaining human relationships between individuals and groups (i.e., it has implications beyond the goods exchanged, as expected of all transactions in nepotistic settings). Obviously though, kin exchange (including gift-giving) is not confined to isolated societies but practiced side-by-side with formal trade in all market societies.

Traditionally, economists have been reluctant to deal with nepotistic exchange in a systematic way. Alfred Marshall, the most influential neoclassical economist at the turn of the nineteenth century, was very explicit about this attitude. "Economics," he noted, "is a study of men as they live and move and think in the ordinary business of life. But it concerns itself chiefly with . . . man's conduct in the business part of his life" (Marshall, 1961:14). Yet even Marshall was willing to entertain a bit of nepotism (in what seems to be a moment of weakness) by approving as adequate for economic inquiry issues like "the distribution of the family income between its various members, the expenses of preparing children for their future career, and the accumulation of wealth to be enjoyed after the death of him by whom it has been earned" (Marshall, 1961:24). Ironically, in so doing, he managed to predict with remarkable accuracy the very topics through which the departure of economics from self-imposed compartmentalization was to take place half a century or so later, notably by economists like Gary S. Becker and Jacob Mincer.[4]

Nepotistic exchange is regulated in large part by universal and quite ancient evolutionary mechanisms: kin selection and sexual selection. It thus falls into an area where human behavior and animal behavior closely overlap. The fact that nepotism has not been as much perfected by humans as by certain species of social insects and colonial animals (given a head start of some 300 million years) should come as no surprise. What has been perfected and exclusively practiced by human beings at an altogether different level of social interaction, is not exchange among closely related conspecifics, but exchange among conspecifics at large: *mercantile exchange*.

Mercantile exchange

On the first day of March in 1834 the HMS *Beagle* was anchored off the coast of Tierra del Fuego, "moored by a beautiful little cove," in Darwin's words, "with her stern not 100 yards from the mountain side." The

---

[4] Rigorous economic inquiry by Becker, Mincer, and their colleagues, has been applied in recent decades to a wide array of "unconventional" topics: discrimination, fertility, parental investment, marriage, home production, crime, addiction, and so on (see, for instance, Becker [1976a,b], and his Nobel Lecture [1993]). Many of these new topics (especially the theory of the family, and to a lesser extent the theory of discrimination) actually deal with *nepotistic exchange*, though the term itself was not used as such.

episode that produced the following entry in his journal took place in the midst of a noisy encounter between the crew and the local inhabitants:

> Both parties laughing, wondering, gaping at each other; we pitying them, for giving us good fish and crabs for rags, etc.; they grasping at the chance of finding people so foolish as to exchange such splendid ornaments for a good supper. ... Some of the Fuegians plainly showed that they had a fair notion of barter. I gave one man a large nail (a most valuable present) without making any signs for a return; but he immediately picked out two fish, and handed them up on the point of his spear. (Darwin, 1945:217; Keynes, 1988:224)

Darwin's colorful account is nicely corroborated by an independent insight made on the same day about the same Fuegian people by the captain of the *Beagle*, Robert FitzRoy. In a laconic entry into the official journal of the expedition he writes: "The 1st of March passed in replenishing our wood and water at a cove, where we had an opportunity of making acquaintance with some *Yapoo Tekeenica* natives, who *seemed not to have met white men before*" (italics added, FitzRoy, 1839:323).[5]

Darwin's encounter with the Fuegian fisherman is a pure and plain example of *mercantile exchange*. The fact that the young Darwin chose in passing to treat it as little more than an anecdotal curiosity, is understandable, but should not detract from its significance. Properly interpreted, nothing in the thousands of pages he subsequently produced marks an attribute that sets humans more sharply apart from other animals. In all fairness to Darwin, as we shall shortly see, he was never in the business of drawing sharp borders between humans and animals. If anything, he was engaged in a life-long struggle against such borders.

Unlike nepotistic exchange, mercantile exchange (or trade) is determined exclusively by the merits of the commodities exchanged.[6] In its

---

[5] Adapted from Keynes (1988:225).

[6] In the sense that the value of the commodity received exceeds the value of commodity relinquished from the separate viewpoint of each trader. Economists would agree with this rule of thumb so long as the term "value" refers to the "total values" or "average values" of the transaction. They will be quick to add, however, that the essential measure that regulates transactions is the "marginal value," which tends eventually to equalize between the traders (assuming divisibility), thus eliminating any initial difference.

idealized form, it depends neither on the identity of the traders nor on any social or biological contingency that has no direct bearing upon the transaction itself. Voluntary transactions under this form of exchange are always settled in explicit commodities rather than in genetic or emotional currency. Mercantile exchange can operate perfectly within or without a concrete marketplace (consider the modern mail-order retail network, or the door-to-door peddlers of earlier times). Unlike the exclusionary patterns of nepotistic exchange, mercantile exchange is all-encompassing. Unrestricted by kinship networks and breeding contingencies, it is wide open to conspecifics at large, to strangers (like Darwin in Tierra del Fuego), and even to anonymous traders. Human beings are probably the only living form engaged in this type of exchange, and are certainly the only living form whose members do so in a systematic and utterly spontaneous manner.

It should be noted that the distinction between nepotistic exchange and mercantile exchange is more than merely a classification device. The two do not mix well with each other in actual transactions. Stigmatization of "nepotism" in business affairs and, conversely, "commercialization" in family affairs, are indicative of considerable resistance to surrendering the necessary levels of coherence in the conduct of both.[7]

Mercantile exchange is a phenomenon of pervasive importance in human affairs. The activity is evident throughout life from child play to bequests, virtually from cradle to grave. It has been practiced throughout history in all known societies, in free markets or (under duress) in black markets. Attempts to limit or uproot it have time and again proven to be futile, no matter what the measure: moral persuasion, political pressure, or outright persecution.[8] Exchange in one form or another has been observed in some of the most inhibiting and precarious human situations such as prisons or POW camps (Redford, 1945). Animosities in the wake of wars may linger for generations, but place few obstacles in the

---

[7] It is probably no coincidence that prostitution, the ultimate mercantile-nepotistic mixed transaction, is also the most austerely stigmatized one in most societies (at least so far as mutually consented-to transactions that have little or no ill-effect on third parties are concerned).

[8] One can scarcely underestimate the capacity of exchange to operate clandestinely. Lessons from the period of (alcoholic) prohibition and the dismal success in the more recent "war on drugs" are only two reminders.

ways of trade.[9] Human traders seem at times to overcome insurmount-
able obstacles, not the least of which is a lack of common language.
Universal body language and gestures, and makeshift pidgin languages of
all kinds, readily fill this gap. From Marco Polo to the crew of HMS *Beagle*,
trade by gesture served the great explorers in their encounters with newly
discovered civilizations (and it seems to serve well modern tourists in the
same predicament). In light of all this relentless indulgence in trade, it is
clear that at some point in its long natural history humankind acquired
a curious disposition and made it its second nature.

Reasonable observers may disagree on the exact origin of this disposition. There should be little disagreement, however, about the role of
trade in modern societies. Exchange (in labor and capital markets) is
practically the only way the vast majority of modern humans earn a living and then (in consumer markets) spend it on the necessities of their
life. What would it be like if the art of exchange had suddenly disap-
peared from the face of the earth? It is clear that civilization as we know
it would cease to exist in a matter of weeks, if not days, and the bulk of
the world's urban population would starve to death in short order.[10]
Whether the final number of survivors would exceed or fall short of the
number to survive a major nuclear confrontation is anybody's guess.

Contemporary human societies literally live by exchange and die by
exchange. The toll claimed by the major famines of the twentieth century – a loss of 100 million lives or so – is almost equal in size to the toll
claimed by the major wars of the century. In the public mind, famines
are associated with droughts or other natural disasters, essentially, with
"acts of God." Some social philosophers may attribute the phenomenon
to Malthusian forces, again of no human fault.[11] However, when the data

---

[9] This is evident in formal provisoes of peace treaties, in the balance of trade between former enemy nations and, most vividly, in widespread and uninhibited exchange between
individuals within hours of cessation of hostilities, if not before.

[10] Consider, for instance, the simple fact that less than 4% of the US labor force is employed
in agriculture – the exclusive source of food for the entire population. This means that,
at any time, more than 96% of the population relies on some form of exchange for subsistence.

[11] Famines (along with war and disease), according to Malthus (1976) are essentially checks
on population growth. This Malthusian theory had great influence on scholars in the
nineteenth century – Darwin and Wallace included. Its main tenets are fairly applicable
to populations of nonhuman organisms. However, so far as human populations are
concerned, the predictive power of this theory, at least in its original version, has been

are examined with some care, it soon becomes apparent that all too often famines are artifacts of human inexpediency: ill-advised or deliberate destruction of local exchange systems that in normal times coordinate the distribution of food and motivate its production (Sen, 1981; Devereux, 1993). In most cases of famine, an adequate supply of foodstuffs was piled up, or the capacity to produce it was in place, either in the midst of starving populations or in nearby regions.[12] In fact, export of food from famine-stricken regions has often continued unabated in spite of local starvation. The movement of food from Ireland to England throughout the Irish famine of the 1840s, and from Bangladesh to India in 1974, are two well-known examples (for other examples, see Sen, 1981).

Despite all these disturbances, at the fundamental level of bilateral transactions, mercantile exchange has proven time and again to be an indestructible ingredient of human behavior.[13] The fact that, in the end, the majority of the population in famine-stricken regions manages somehow to survive is the best indication to this effect. Evidently, some elements of the distribution system always remain intact. In this capacity, exchange seems to provide an invisible safety net to most people, most of the time, but does not necessarily guarantee the survival or well-being of any particular individual or group. In other words, it gives all the signs of a structure produced by natural selection.

Tentative conclusions

The following points should summarize the main conclusions from the discussion thus far. First, concerning basic patterns of either symbiotic exchange or nepotistic exchange, we find little difference *in kind* between human beings and animals. There are obvious differences in

---

largely discredited by recent trends. Contrary to the Malthusian scenario, food supply has never fallen significantly behind the world population. For most areas of the world – with the exception of parts of Africa – the expansion in food supply has been comparable to, or faster than, population growth (Sen, 1981:7).

[12] This is true, for instance, for the famines in Russia (1934), Bangladesh (1943), China (1958–61), Ethiopia (in the 1980s), and more recently in Somalia and other African countries.

[13] Consider the lesson learned from the European experience with hyperinflation in the post World War I era. Networks of exchange remained largely intact and, against all odds, did not cease to function despite a catastrophic collapse of the entire monetary system.

form, style, and degree, but these differences do not exceed the variation one expects to find across different species. On the other hand, there is no instance in the animal kingdom that nearly resembles the basic pattern and function of mercantile exchange. No species other than humankind relies on free exchange among conspecifics-at-large for its subsistence, no other species is engaged in trade based purely on the merits of the commodities exchanged, and for that matter, no other species settles transactions – as Adam Smith used to say – "by treaty, by barter, and by purchase." With this understanding, his salient observation that "the propensity to truck, barter, and exchange one thing for another" is a human predisposition "to be found in no other race of animals," should be self-evident.

3   Classical economics and classical Darwinism

Darwin and the Scottish economists: The first point of junction

The fundamental economic problem of human evolution

In view of the discussion thus far, the issue is no longer the mere *existence* of a human predisposition to exchange, but its evolutionary *origin*.[1] Unsure of this origin, Adam Smith himself acknowledged (and deflected) in passing an intriguing question:

> Whether this propensity [i.e., to exchange] be one of those *original principles in human nature*, of which no further account can be given; or whether, as seems more probable, it be the necessary *consequence of the faculties of reason and speech*, it belongs not to our present subject to enquire. (1976:17, italics added)

The inception of modern economics was thus accompanied (in 1776) by an evolutionary question preceding by nearly a century the Darwinian notion of evolution itself. Adam Smith probably agonized over this question, though as it seems, had the good sense to abort it in due course. Neither he nor his pre-Darwinian readers could fully comprehend the question, let alone conceive a sensible answer to it. With the advantage of hindsight, however, the question seems to present a challenge of the highest order to the modern study of human evolution: Was exchange an early agent of human evolution, or was it merely a late by-product of previously evolved "faculties of reason and speech"?

With the publication of *The Descent of Man* a century or so later, Darwin had at his disposal a fairly mature notion of evolution applicable to human affairs. At long last a frame of reference for contemplating the dilemma posed by Adam Smith came into being (though a solution was

---

[1] When Adam Smith, like most economists, uses the term exchange (without a modifier!) he means, mercantile exchange. In the interest of brevity I will henceforth follow the same convention (except where the shorthand invites confusion).

not forthcoming). Darwin made a problematic distinction between the class of evolutionary phenomena associated with "the advancement of man from a semi-human condition to that of the modern savage" and the class of phenomena associated merely with "the action of natural selection on civilized nations" (1874:136). One can only wonder in which class he would place the phenomenon of human exchange. As far as I am aware, Darwin himself provided no clue to his position on the issue – except, perhaps, for one brief comment which I will discuss at a later point . However, the relevant tenets of his thinking and the principles he established shoulder to shoulder with Alfred Russel Wallace do, indeed, have some bearing on this particular issue.

Darwin's self-restraint

Exchange and other niceties of human subsistence play little or no role in *The Descent*, let alone in other written works by Darwin. For whatever it was worth, the influence of economics on Darwin was inspired by abstract ideas rather than by operational applications.

The core insight that links Charles Darwin to Adam Smith is a common recognition of the possibility of design-without-a-designer: the plausibility of spontaneous order. Of course, the clockworks differ and the watchmaker is not the same. There is no reason to turn a blind eye to crucial differences. For instance, to Darwin the individuals are evolving, but the niche that a population occupies in the environment is (more or less) fixed. To Adam Smith the individuals are fixed, but the niche (the economy) is evolving. Similarly, to Darwin the individuals are in a ceaseless state of competition *against* each other. To Smith, the individuals compete (most of the time) for trading partners; i.e., they are largely in a state of competition *for* rather than *against* each other (for other subtle distinctions between the two paradigms, see Gordon, 1989).

Yet, the two intellectual enterprises – "natural selection" and the "invisible hand" – are driven essentially by the same generic idea of competition and optimization (at the level of individuals) and equilibrium (at the level of populations). This metatheoretical connection, combined with the Malthusian theory of population, lends support to a general evaluation pointedly expressed by Stephen J. Gould:

> The sources [of ideas that most influenced Darwin] were many, various, and exceedingly complex. No two experts would present

the same list with the same ranking. But all would agree that two Scottish economists of the generation just before Darwin played a dominant role: Thomas Malthus and the great Adam Smith himself. (1993:148)

There is little doubt about Thomas Malthus's influence on Darwin (and for that matter, on Wallace). But the full extent of Adam Smith's influence, as indicated, is still a matter of some dispute among the experts.[2] This is not the place to discuss the extent to which Darwin was influenced by, or even aware of, economic phenomena that lie wholly outside biology. The essential point here is Darwin's omission – or near omission – of *exchange*, a phenomenon almost as indigenous to biology as it is to economics.

Darwin did not single out human exchange for exclusive neglect. His omissions included issues far more important to the general theory of evolution. The origin of life, the specific ancestry of man, and (for the most part) the human mind – were all omitted by conscious restraint rather than by oversight or lack of interest on Darwin's part.[3] None of these issues, it should be noted, has been fully resolved to this day. With the benefit afforded by hindsight it is safe to argue that Darwin's restraint was well warranted. Had he chosen to take any of these issues head-on, his larger enterprise may have never reached fruition in a single generation – especially in a generation such as his own. *The Descent of Man*, the first part of the work anyway, was produced primarily for readers of that generation.

The Descent is not a comprehensive or general outline of human evolution, nor was it intended as such. "The sole object of this work," Darwin emphasized, "is to consider, firstly, whether man, like every other species, is descended from some preexisting form" (1874:2). An attempt to establish and advance the idea of evolutionary continuity from animals to human beings is evident in his choice of topics and exposition.

---

[2] Schweber (1978, 1980), Gordon (1989), Schabas (1994) – to name but a few.

[3] Darwin's reference to these three items in a handwritten paragraph entitled "Omissions" is preceded by the remark "Extent of my theory – having nothing to do with ..." (as cited from the *Pencil Outline* in Schweber, 1978:325). At some point, earlier on, he was actually contemplating avoiding the subject of humans altogether in his evolutionary theory while, in his own words, considering it "the highest and most interesting problem for the naturalist" (Eiseley, 1961:256).

Darwin seems to go to great lengths to show that no human attribute is so peculiar that some nuance or vestige of it cannot be found in animals. To illustrate, consider his attempt to lay to rest the long-standing question of the missing human tail. "According to a popular impression," he points out, "the absence of a tail is eminently distinctive of man; but as those apes which come nearest to him are destitute of this organ, its disappearance does not relate exclusively to man" (1874:58). Indeed, *The Descent* treats human evolution with an emphasis on anatomy and aspects of sexual selection – two areas in which characteristics common to humans and animals are most easily observed. A secondary emphasis on morality and on emotions, especially emotional expressions, focuses on instincts and the neuro-physiological processes that accompany these mental phenomena, rather than on cognitive implications.[4] The same approach is taken, and elaborated upon, in *The Expression of the Emotions in Man and Animals*, Darwin's sequel to *The Descent*. To establish evolutionary continuity Darwin had to rely on humanlike characteristics observed in animals (or, equivalently, on animal-like characteristics in humans). Early evolutionists, in Loren Eiseley's view, unconsciously stressed such characteristics almost to the exclusion of other factors:

> Man, theologically, had for so long been accorded a special and supernatural place in creation that the evolutionists, in striving to carry their point that he was intimately related to the rest of the world of life, sought to emphasize those characteristics which particularly revealed our humble origins. (1961:288)

The fact that humanlike characteristics in animals were intriguing to Darwin and some of his contemporaries (if only for their anthropocentric mindset), and offensive to many others (for the same reason), made Darwin's task especially challenging. Exchange and other aspects of human subsistence – areas of difference rather than similarity between humans and animals – had to fall by the wayside.

The main difficulty with Darwin's argument of continuity from animal to man, apparently the only one with lasting effect, was directly

---

[4] Studies dealing with localization of brain functions, starting with the work of Paul Broca (1878) in general, and followed in particular by the works of Papez (1937) and MacLean (1973), now largely associate these processes with the *limbic system*: a site in the brain that humans share with other animals.

associated with the human mind. As I have already indicated, the topic was undertaken with great reluctance. Some critics consider Darwin's overall treatment of this particular topic to be highly anecdotal, and he himself considered it (perhaps, intended it to be) incomplete (1874:129). The least impressed by Darwin's approach to the evolution of the human mind was none other than his loyal partner, the codiscoverer of natural selection, Alfred Russel Wallace himself. In a rare outburst of disapproval, he pointed out that

> . . . to prove continuity and the progressive development of the intellectual and moral faculties from animal to man, is not the same as proving that these faculties have been developed by natural selection; and this last is what Mr. Darwin has hardly attempted, although to support his theory it was absolutely essential to prove it. (1889:463)

Wallace was in a unique position to make this observation. Of all his contemporaries, within and without the Darwinian circle, Wallace was the one destined to sense and faithfully divulge the single most embarrassing difficulty confronting his own (and Darwin's) theory.

In coping with the question of the human mind, Wallace, for the most part, was far ahead of his own time. Though little in the way of paleontological findings was available, it was already clear to him and to most experts of his day that the expansion of the human brain represents an unprecedented example of plasticity in geological time. No single major organ has been observed to grow at nearly the same rapid rate (in proportion to the body size of its carrier). However, what was clear to Wallace, but not to most of his contemporaries, was that evolutionary dynamics on the scale of geological time do not carry over onto the scale of recent historical time. Variables in the long run are often short-run constants. If the human brain and its mental derivatives were produced by natural selection then they should be fixed on the average (and limited in variance) going back to prehistoric members of the species and, by implication, across existing human societies and races. With some important reservations, to be discussed later, Wallace accepted this state of affairs as an empirical reality and as a necessary condition for the action of natural selection. For him the faculties of the human mind (latent if not active), like any other species-specific characteristic, were no subject for development ladders over recent historical events or across existing people. (Assignment of such ladders, it seems, was a popular pre-occupation among nineteenth-century scholars.) He ascribed any defi-ciency in use or performance of mental faculties, if actually observed on the average in a (large) population, to insufficient "means" or "incite-ments" in the immediate society – that is, to instrumental latency rather than to evolutionary retardation. It comes as no surprise, as we will shortly see, that the existence of an agent unaccounted for by the pre-vailing description of human evolution did not escape Wallace's keen insight. It was probably the same insight which led the ninety-year-old Wallace, in the last year of his life, to take a solitary position in rejecting the significance of the newly found Piltdown man. It took his younger colleagues 37 more years to reach the same conclusion (when in the end the finding proved to be a hoax).

Darwin's principle of utility: The second point of junction

Any attempt to account for human exchange is absent from the writings of Wallace as much as from the writings of Darwin. But with Wallace the absence is more conspicuous simply because he was ready and willing to ponder the evolution of the human mind without the inhibitions typical of Darwin. There are other differences. Unlike Darwin who was first and foremost an advocate of descent and evolutionary continuity, Wallace was the guardian of natural selection. Certain conflicts between the partners, however rare, were inescapable.

In his autobiography, Wallace lists four areas of conflict between him-self and Darwin (Wallace, 1908:236–37). Only one of these rare clashes is of direct relevance to the present discussion: the disagreement about the emergence of human intelligence. The legacy of Wallace was eventually badly marked by this very conflict.[5] At the bottom of this conflict lies a fundamental rule of natural selection: the *principle of utility*. In Wallace's own words,

---

[5] It is interesting to note that in two of the three remaining issues, the disagreements about sexual selection and about inheritance of acquired characteristics, Wallace was firmly on the defense of natural selection – strictly interpreted – whereas Darwin was in partial retreat. The fourth disagreement dealt with the mechanism of intercontinental seed dispersal as related to arctic and mountain flora.

> ... none of my differences of opinion from Darwin imply any real
> divergence as to the overwhelming importance of the great
> principle of natural selection, while in several directions I believe
> that I have extended and strengthened it. The principle of
> "utility," which is one of its chief foundation-stones, I have always
> advocated unreservedly ... [and] extended its range. Hence it is
> that some of my critics declare that I am more Darwinian than
> Darwin himself, and in this, I admit, they are not far wrong.
> (1908:237)

Natural selection, according to the principle of utility, can produce neither a structure harmful to an organism, nor a structure that is of
greater perfection than necessary for an organism at a given stage in its
evolutionary history. Neither overdesign nor foresight are admissible
under natural selection. The term *Darwin's principle of utility* was probably
coined by Wallace himself (economists who use "utility" in a slightly different sense would probably prefer here something like "parsimony" or
"myopic efficiency"). Darwin implied that a single counter-example to
this rule would be fatal to his theory (Darwin, 1964:200–2).[6]

Barely a decade had passed since the publication of *The Origin* when
Wallace first sounded the alarm bells (1869, 1870). Equipped with the
principle of utility, he called into question the applicability of natural
selection to the evolution of the human intellect. The human brain, its
higher mental faculties (the capacity for mathematics, music, poetry,
etc.) and certain physical characteristics (the human hand) – all conveyed
to Wallace attributes of greater perfection than was necessary for survival at the time in which they evolved. Wallace soon arrived (perhaps
too soon) at the drastic conclusion that humankind's history cannot be
reconstructed purely in terms of natural selection. In the end, he relegated the evolution of the human mind to "some agency other than natural selection, and analogous to that which first produced organic life."
After some modifications, he confined his stipulation only to "moral and
intellectual qualities" and not to "physical forms." Ironically, in his

---

[6] Perhaps too few authors in the Darwinian tradition, especially in the popular periphery
of the literature, have paid heed to this rule by subjecting all their applications to its critical scrutiny. This should not be taken as an indication that the rule is no longer relevant.
A vivid reminder can be found in a chapter from a recent work by one of Darwin's best
known disciples (Dawkins [1995]). The chapter is entitled "God"s utility function."

attempt to shield natural selection he was willing partly to exempt human evolution from it.

Darwin's deep and understandable anxiety over his partner's unexpected willingness to throw away the baby (continuity) and keep the bathwater (natural selection) is implicit between the lines of the *Descent*. More explicit expressions to this effect are evident in private correspondences: "I hope," he wrote to Wallace when he first learned of his intention partly to exempt humans from the grip of natural selection "you have not murdered too completely your own and my child." Overwhelmed by the sound logic of Wallace's dilemma, or (more likely) bewildered by his solution at the borderline of "heresy," Darwin's most prominent backers (like T. H. Huxley and Asa Gray) were slow to respond in favor of Darwin on this issue. Some (like Charles Lyell) actually showed an inclination to side with Wallace (Bowlby, 1990:394). Attempts by less prominent supporters (e.g., Write, 1870), and by Darwin himself, to resolve the issue within the range of natural continuity produced some interesting specific explanations, but evidently failed to produce an overall plausible explanation capable of removing all doubts. "Any Darwinian has to admit," writes one of the most careful modern students of classical Darwinism in reference to the controversy surrounding Wallace's approach to human evolution, "that we humans present some awkward cases for natural selection" (Cronin, 1991:357). Sure enough, a prominent Darwinian like Edward O. Wilson is unafraid to acknowledge the difficulty in the open:

> Natural selection, in short, does not anticipate future needs. But this principle, while explaining so much so well, presents a difficulty. If the principle is universally true, how did natural selection prepare the mind for civilization before civilization existed! *That is the great mystery of human evolution*: how to account for calculus and Mozart. (1998:48, italics added)

Wallace's dilemma apparently still looms large in the minds of Darwinians and, next perhaps only to the missing gaps in the fossil record, it stands as one of the most important remaining enigmas in the modern study of human evolution.

Separate approaches to a common puzzle

In its most fundamental sense, the dilemma that Wallace failed to resolve neatly overlaps the very dilemma Adam Smith was unable to resolve a century earlier. Adam Smith, as we saw, was missing a clear concept of evolution. Alfred Wallace, on the other hand, was missing a clear concept of human exchange, or at least was unwilling to tackle its subtleties. In this respect the two are paradoxically connected by separate missing pieces to a common unresolved puzzle.

Adam Smith went to great lengths to emphasize the "utility" of exchange (the advantage it confers onto its participants through division of labor), but was ambivalent about whether exchange is a consequence of the faculties of reason (and speech), or whether it is an independent original cause. The ambiguity between these two alternatives is removed, however, when Wallace's argument is brought to bear on the issue. That exchange is a consequence of the human faculties of reason is clearly inadmissable under Darwin's principle of utility (as interpreted by Wallace). Thus, by elimination, the second alternative is more appealing from the standpoint of natural selection. The implication is that exchange was largely an independent agent of evolution and, as such, was more the cause of the faculties of reason and speech than their effect. I do not know if Wallace was altogether aware of the possibility that human exchange could have played an early and distinct evolutionary role in this capacity. An attempt to explore such a possibility might have saved him, however, an awkward detour in his long and unrewarding quest for the elusive origins of human intelligence.

Bootstrap encephalization

Whatever their source, the faculties of reason, speech, and intelligence in general, can scarcely reach levels of perfection deemed unnecessary under market exchange. The mental skills of traders that make strategic exchanges with genetically unrelated members of their own species are constantly challenged in an adaptive sense. The benefits from trade gained by exchanging parties depend not only on their *absolute* level of intelligence, but also on their *relative* level of intelligence with respect to each other. A trader endowed with a relatively inferior level of intelligence is in a position of *comparative* disadvantage. Natural selection,

Wallace would be the first to insist, leaves little room for such disadvantages. If one trader gets a little smarter, others must follow suit. Once it has been started, the evolutionary cycle feeds on itself and keeps escalating from generation to generation until it reaches an equilibrium at a point which is not necessarily optimal from the standpoint of all traders, taken as a group. The result is a self-reinforcing process of competition among conspecifics analogous to an escalating arms race among nations. The process is not unique to human beings. As we shall see in a subsequent chapter, the giraffe (the tallest living animal) and the *Sequoia* tree (the tallest living plant), both have probably acquired their imposing, almost absurd, heights through similar runaway processes of self-reinforcing competition for subsistence among conspecifics. What human beings have acquired through such a bootstrap evolutionary process, or through something close enough, is a brain of unprecedented proportion and complexity. If this grossly enlarged organ, to which we attribute much of our mental capacity, was indeed produced in response to self-reinforcing selection pressures, then it should represent neither overdesign nor foresight on the part of natural selection. Just by recognizing this possibility one is already more Wallacean than Wallace himself.

Diversity of human nature: The third point of junction

The single most comprehensive and probably the most important piece of work in nineteenth-century economics, *Principles of Economics* by Alfred Marshall, was published in 1889 – almost a decade after Darwin's death. Marshall was already in a position to make a preliminary evaluation of the impact of Darwinism on the social sciences and on economics in particular. As he saw it,

> At last the speculations of biology made a great stride forward: its discoveries fascinated the attention of the world as those of physics had done in earlier years; and there was a marked change in tone of the moral and historical sciences. Economics has shared in the general movement; and is getting to pay every year a greater attention to the pliability of human nature, and to the way in which the character of man affects and is affected by the prevalent

> methods of the production, distribution and consumption of
> wealth. (1961:764)

Economics, behavioristically considered, relies more than any other field in the social sciences on human pliability: that is, on choice, on the ability to make choices, and – to some extent – even on the human tendency to undergo certain changes in character and personality in response to available choices (or their absence). Pliability of human nature, as we shall shortly see, was a stepping-stone in Adam Smith's approach to human diversity. The wide range and high level of variation in human abilities, compared with those of animals, have mystified many other insightful observers outside of economics and throughout history – from ancient Greek philosophers to modern geneticists and evolutionary biologists. But to none was the issue of variation in human characteristics more critical than to Wallace. Once again, the names of Smith and Wallace are interwoven under yet another issue. The issue, this time around, is the ancient problem of human diversity.

Wallace's "independent proof"

For Wallace, the issue of human diversity closely revolved around the concept of species-specific characters (i.e., peculiarities determined by genes that distinguish one species from another). Such characters are pivotal to taxonomic classification, but to classical Darwinism they always conveyed an additional meaning. Both Darwin and Wallace viewed them as observable indicators against which some predictions of natural selection could be tested (especially predictions about the fine balance between variability and stability of adaptive structures). If a structure or a function was produced by natural selection, it could be expected to follow a rigid law of stability expressed and empirically quantified by Wallace as follows:

> From its very nature [the process of natural selection] can act only
> on useful or hurtful characteristics, eliminating the latter and
> keeping up the former to a fairly general level of efficiency. Hence
> it necessarily follows that the characters developed by its means
> will be present in all the individuals of a species, and, though
> varying, will not vary very widely from a common standard. The
> amount of variation [estimated in his own empirical studies is]

about one-fifth or one-sixth of the mean value – that is, if the
mean value were taken at 100, the variation would reach from 80
to 120, or somewhat more, if very large numbers were compared.
(1889:469)

Moreover, according to his observations and calculations, the proportion
of specimens that reach extreme or nearly extreme performance (or size)
in a given character is expected to be from 5% to 10% of the population
examined (Wallace, 1889:81). Relevant characteristics of animals appear,
at least in Wallace's estimations, to be in line with this law – as are probably many human characteristics. What he found to be in defiance of
this law was, once again, the human intellect and its higher faculties.
The proportion of individuals in any human population that possess the
genius of composing a fine piece of music, or the genius of coming up
with a deep and elegant theorem in mathematics, seems to fall far short
of these expectations. Wallace actually surveyed music and mathematical masters (in one of the "great public schools") as to their students'
prospects of reaching such levels of virtuosity. The response was quite
dismal. Only 1 in 100 or so students was found to have "real or decided
music talent," and, curiously enough, exactly the same small proportion
(but not necessarily the same students) were found to possess "the natural faculty which renders it possible for them ever to rank high as mathematicians, to take any pleasure in it, or to do any original mathematical work" (1889:470–71). Such small proportions imply a sizable range
and variance in human aptitudes which seemed, at least to Wallace, to
exceed by far the regularities of natural selection. According to his interpretation, these observations provide an independent confirmation in
support of his controversial stand on the human intellect; namely, that
it could not have developed under the law of natural selection. He called
it the "independent proof" (1889:469–72).

Any attempt to critically evaluate the "independent proof" should not
rule out, I think, the possibility that variation itself is selected under
phenotypic plasticity; that is, an adaptive characteristic may include
variation (at the phenotypic level) as part of the adaptation itself. In all
fairness to Wallace, such a critical evaluation should also recognize the
fact that the distinction between phenotypic plasticity and genetic
variation was not entirely clear in his time. Under phenotypic plasticity,

a genotype may develop different reversible states for a given character, or a population may develop different frequencies from within a given set of genotypes. In either case, a large degree of phenotypic plasticity entails no more genetic instability than, say, a large degree of sexual dimorphism. With this understanding, we may conclude that what the "independent proof" actually means is not a deviation from the regularities of natural selection, but the possibility that the human mind is subject to a remarkable level of phenotypic plasticity. Indeed, as we shall shortly see, the existence of plasticity over a wide range of human mental and vocational faculties is actually implied, rather than denied, by the very action of natural selection.

The blunder of Epimetheus

The heterogeneity (or inequality) in human native abilities was a source of curiosity and concern to insightful observers long before Wallace. It was certainly on the minds of some ancient Greek scholars as they reflected on nature and human nature. As it seems from the Great Speech of Protagoras, in the Platonic dialogue *Protagoras*, human diversity is part of an evolutionary fiasco (Plato 1978, 320c8–328d2).[7] The narrative is best understood when stripped of its mythical language (as the ancient author in his opening remarks implicitly urges the reader to do). In the idealized Protagorian "economy of nature" all animals were created morphologically and ecologically perfectly adapted "with the view of preventing any race from becoming extinct." The sole error of genesis and, so to speak, the only endangered species is the human being – a hairless, barefoot, helpless creature without means of protection from predators or means of subsistence. A belated compensating adaptation, the stolen *Promethean fire* (i.e., technology, along with the mental and vocational faculties that enable it, but as yet without political wisdom) turns out to be part of the problem as much as part of the solution. Unlike the homogeneous adaptations uniformly bestowed on members of other species, the *Promethean fire* apparently was unevenly and unequally distributed among members of the human species.

---

[7] The ancient Greeks had, of course, a number of free-spirited competing theories of evolution including, according to Osborn (1929), an embryonic theory of natural selection (advanced by Empedocles and rejected by Aristotle).

Consequently, left to their own devices, human beings could not survive in isolation, and human societies could not survive without institutions that coordinate the action of inherently dissimilar members. The capacity to form and maintain such man-made political and economic institutions was eventually delivered (by Hermes) in the form of morality, a sense of justice, and "the art of government." Unlike Prometheus, Hermes made sure – even double-checked with Zeus himself – that this redeeming adaptation would be fairly evenly distributed among all human beings. The *logos* behind this evolutionary *mythos* is, among other things, a psychological defense of democracy. In spite of all the natural inequality in their ability to master technical arts, all citizens can and should partake of the political art.

For all the difference in substance and style that set an ancient philosopher apart from a Victorian naturalist, there is a certain similarity between Protagoras (probably the man and certainly the dialogue) and Wallace's basic approach to human diversity.[8] Both use (intraspecific) variation in animals as a point of reference for human diversity. By this standard, both find excess diversity in certain human traits but not in others. Excess diversity in the "mechanical arts" is emphasized in the Protagorian dialogue. Wallace found it, close enough, in the "higher faculties of the human mind."

Moving from nature to politics, human diversity was still high on the mind of the ancient Greeks. Close attention to the problem of inequality in human native abilities is evident in the way they contemplated and practiced statecraft and public life. Thus, for instance, the ideal state in Plato's *Republic* – a segregated three-class structure ruled by a philosopher king – is a masterpiece of social engineering unmistakably designed to address this very problem. The Athenian model of democracy under the regime of Pericles (and the influence of his friend Protagoras) addresses the same problem but in quite a different, perhaps more appealing, way. In its ideal form it was conceived as a community of *citizens with dissimilar qualities and abilities*, but with equal duties, equal rights before the law, and equal opportunities (Loenen, 1941:13–14). On a more fundamental level their solution to the "problem" of human

---

[8] Though the dialogue itself was written by Plato, the most recent consensus among experts is that the Great Speech of Protagoras itself is, for the most part, an authentic reproduction of one of Protagoras' works (see, for instance, Schiappa, 1991:146–148).

diversity relied on two economic institutions: division of labor and exchange.

In practice, and purely as principles of action, division of labor and exchange were perfected in the ancient world before the Hellenic era. For a millennium or so the Phoenicians had dominated the Mediterranean precisely with the aid of these two principles. Their Greek neighbors did not trail far behind. The overall importance of division of labor and the degree of specialization in Greek societies is evident in historical accounts. The following by Xenophon (430–355 BC) refers to shoemaking in "large cities."

> one man makes shoes for men, another for women, there are places even where one man earns a living just by mending shoes, another by cutting them out, another just by sewing the uppers together, while there is another who performs none of these operations but assembles the parts. (*Cyropaedia* 8.2.5, reprinted in Finley, 1973:135)

A full appreciation of this phenomenon would have to wait for its interpretation by Adam Smith (or its implementation by Henry Ford) millennia later.

On their part, the Greek thinkers had largely associated division of labor and exchange with a solution to the problem of human native diversity. Exploring every avenue through which a society of dissimilar members can mitigate the ill effects of its own diversity, they came up with a practical arrangement: Each member needs simply to specialize in a narrowly defined activity, one in which his or her peculiar shortcomings are minimized. Thus, for instance, in Book II of *The Republic* (370a–370c), Plato reminds us that

> ... there are diversities of natures among us which are adapted to different occupations. ... And if so, we must infer that all things are produced more plentifully and easily and of a better quality when one man does one thing which is natural to him and does it at the right time, and is free of other pursuits.

The fact remains, however, that narrowly specialized producers are never completely self-sufficient. They mostly work for the necessities of others, and they rely on others for most of their own necessities. Consequently, they cannot survive without exchange. "And they exchange goods with one another, both giving and taking, under the idea that exchange is for their own good" (*ibid.*, 369c). Exchange, according to this ancient chain of reasoning, is a prerequisite of division of labor as much as division of labor is a prerequisite of sheer existence – assuming, to begin with, that human beings are created unequal.

Notably missing from this chain of reasoning is the silver lining: the recognition that division of labor is in itself a productive agent of great advantage, regardless of prior diversity in native abilities among the participants. Consequently, the main implication was kept at bay. So long as division of labor was viewed as a device largely intended to compensate for the ill effects of the diversity of human nature the advantage it affords was grossly underrated. This narrow view of division of labor fits neatly into the mindset of Hellenic philosophers and their need to explain the place of man in the cosmos (or the place of slaves in their own backyard), but had little to do with the real world as Adam Smith saw it.

The second fundamental problem

We have already had occasion to explore the first economic fundamental problem of human evolution raised (without being fully aware of it as such) by Adam Smith in reference to exchange and the emergence of the "faculties of reason." The second fundamental problem refers to human diversity. Obviously, Smith was no longer present when Wallace dealt with these two problems using his own devices (i.e., the principle of utility and the "independent proof"). However, he was surely familiar with the philosophical and practical thinking of the ancient Greeks as outlined above. In his own work, Smith adapted some of their more sensible economic tenets (such as the nexus between exchange and division of labor), but never really used them as touchstones for his own ideas. On the issue of human diversity his departure from the Greek position was clear:

> The difference of natural talents in different men is, in reality, much less than we are aware of; and the very different genius which appears to distinguish men of different professions, when grown up to maturity, is not upon many occasions so much the cause, as the effect of the division of labour. (1976:19)

In other words, he turned the Greek paradigm on its head. At the root of this assertion is again a fundamental evolutionary question: is diversity an agent of the division of labor or merely a byproduct of it? The question is a fairly straightforward (empirical) issue if conceived on the scale of a single generation. It gets more challenging, and more captivating, if conceived on the scale of evolutionary time (when the advantages of division of labor can operate on diversity not only through incentives to specialize but also through selection).

The assertion just cited was not made in the abstract. Adam Smith was the first to grasp the role of the division of labor in society in its fullest extent – as a universal agent of productivity and a major source of wealth of nations. His most compelling observation to this effect was made in a tiny pin factory (with ten workers). Obviously, it should come as no surprise that ten pinmakers doing each only one minute task can turn out more than ten times the number of pins a pinmaker working alone can possibly make. But *more* by how much? By Adam Smith's account, output in the factory under his observation surpassed at least by a factor of 240 to 1(!) the output of ten workers operating apart. This remarkable increase in productivity afforded by division of labor had apparently little to do with the initial distribution of natural talents among the workers.

Exceeding any expectations of a casual observer, the advantages afforded by division of labor are large, and the larger they are, the greater are the incentives that drive workers to specialize – assuming, of course, that free exchange or some other system of redistribution is in place. (In the absence of such a system the benefits of division of labor could not be internalized by individual agents and the necessary incentives would not be created.) Consequently, any society, even one initially composed of perfect clones, would eventually display intense diversity of (acquired) abilities and talents. This is at least the impression one gets from Adam Smith's further clarification of his position: "The difference between the most dissimilar characters, between a philosopher and a common street porter, for example, seems to arise not so much from nature, as from habit, custom, and education" (1976:19). In remarks like this he seems to cross a fine line between nature and nurture, but does not burn any bridges behind. At no point does he deny the existence of variation in human native abilities and talents. It is clear, however, that in his final account of human diversity he considers such natural variations to be *empirically* not nearly as consequential as the variation acquired through training and specialization.

Adam Smith could not possibly appreciate the full impact of division of labor as it operates from generation to generation via selection. Ongoing incentives in response to the benefits of exchange and division of labor can explain the *motivation* to specialize, but not the *capacity* to acquire specialized training. Smith was thus poorly equipped to provide a plausible explanation for the ease – often enthusiasm – with which human beings specialize in exceedingly narrow productive activities (compared with the general resistance to specialize in consumption). In a pre-Darwinian world, even Adam Smith was yet in no position to grasp the *evolutionary* implications of the advantage afforded by division of labor.

4   Evolutionary implications of division of labor

Division of labor can evolve (i.e., be selected for) only if all its structures and partaking entities are properly exposed to natural selection. By implication, division of labor can be anticipated to readily evolve only if it occurs at a level of organization *strictly below* the unit of selection. For instance, the division of labor among cells and organs of a single healthy organism is possible largely because the unit of selection typically rests at the level of the organism, if not above it. By its very nature, division of labor is an interaction among two or more partly independent entities that have to share, somehow, the costs and benefits of their common endeavor. They need a system of redistribution. This is provided and best regulated by the designated unit of selection on behalf (and from above) all its constituents. For the same reason, division of labor can evolve only with great difficulty at or above the level of organization of the unit of selection. Ordinarily it does not successfully occur at such levels, and those rare instances where it does are invariably associated with evolutionary innovations of great interest. These innovations are the main subject to be discussed in the present chapter.

Some aspects of division of labor are transitory. They may rise in one generation only to fade away in another depending on the changing environment and, in human society, on the rise and fall of contemporaneous technologies (an example is provided by the ever-changing sexual division of labor in human society – to be discussed in a subsequent section). Notwithstanding such transient manifestations, the fundamental principles of division of labor are universal and timeless. The ultimate advantage afforded by division of labor – production of greater quantities at higher quality with less input over a shorter course of time – is generically applicable. This advantage was as germane to hunting-gathering and to early agriculture as it was to an eighteenth-century Scottish pin factory, or to an early twentieth-century assembly line, and is as germane today in the networks of modern industry as it is in a hive of bees. In all cases, however, the benefits of division of labor in society depend in large measure on two conditions: on the capacity for special-ization and differentiation in function and form, and on the capacity to mobilize large contingencies of constituents in conflict-free coordinated action (i.e., on the capacity for recombination). For the reasons just out-lined, both capacities in their ideal form are readily observed in embry-onic cells and tissues but are hard to come by in societies of free living organisms. Consider first the capacity for specialization.

The capacity for specialization and differentiation

The capacity for specialization can greatly enhance the advantage of divi-sion of labor. By its very nature, specialization displays endless peculiar-ities that differ from species to species (each according to the tasks important to its survival). All these peculiarities, however, are governed by a common principle: a high degree of task-oriented phenotypic plas-ticity either in morphology or in behavior, or in both. The degree of phe-notypic plasticity is elevated precisely because diversity is selected for, and diversity is selected precisely because it maximizes the benefits of division of labor (through complementarity and synergy). The essential implication, other things being equal, is that the capacity for specializa-tion on the part of individual members of a species is expected to be directly related to the extent of division of labor among them. Whether the necessary degree of specialization is achieved through the capacity for learning or through instinct, or simply through the capacity to pro-duce a wide spectrum of (fixed) specialized offspring, is a question of sec-ondary importance from a broader evolutionary point of view. These modes of specialization and differentiation are not mutually exclusive. All three are fairly enhanced in humans and, to different degrees, in other social species.

The action of natural selection with reference to division of labor pro-vides therefore a plausible explanation, not only for the human propen-sity to exchange, but also for the exceedingly variant faculties of the human mind, the very question that for centuries troubled separate observers from the ancient Greeks to Smith and Wallace, and beyond. Still, one may actually wonder why the plasticity of the human mind does not carry over to the human body. Strictly speaking, the human body displays little out of the ordinary in morphological plasticity. A closer look reveals, however, a large amount of *functional* plasticity equivalent if not superior to morphological plasticity. Almost all the functional plasticity that could be desired by an organism is marshaled by a single organ: the human hand. Supported by finely tuned sensory receptors (of touch and stereoscopic color vision) and equipped with tools (stones, sticks, strings, and skins – for beginners) the human hand comprises all-in-one a countless number of functionally distinct organs. The ability of the human hand to dress and undress its carrier in a matter of minutes, if not seconds, is only one among many adaptations in this open-ended catalogue of task oriented plasticity. This humble ability is probably worth, however, millions of years of awkward anatomical and physiological adaptations. Properly dressed, a human being can move through all the seasons, from the equator to the poles – and recently even walk in outer space – all without changing body size or shape to meet graceless ratios of surface to volume, without growing clumsy unremovable furs, without secreting oily substances to the surface of the skin, without adding antifreeze to plasma, and without seasonal migration or hibernation – to mention but some inconveniences of an unclad organism. One can go on and, starting with the question of handedness (Box 4.1), fill a book with adaptations associated with the human hand

---

Box 4.1 The human hand: a single organ or two?
From the viewpoint of task-oriented plasticity, the human hand represents two distinct and fairly specialized organs: the right hand and left hand. The human hands seem to be perfectly adapted to complement each other, but only imperfectly adapted to substitute each other, just as expected of two productive agents under a flexible system of division of labor. Ambidexterity, the ability to use both hands with equal skill, is not a particularly appealing proposition either to an accomplished violinist or to a mother who (typically) cradles her baby on the left-hand side. It obviously makes little sense to the bioengineering of tool use. From this point of view, handedness, asymmetry, and the general lateral orientation of the human body and brain are, arguably, further manifestations of task-oriented plasticity.

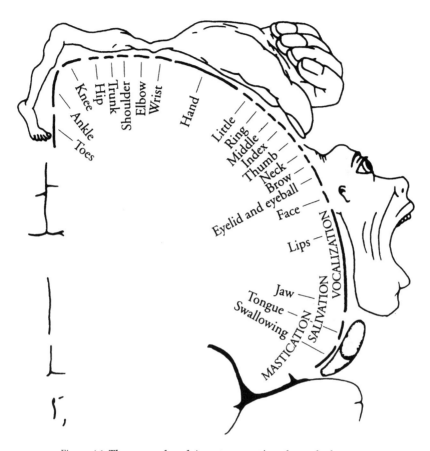

Figure 4.1 **The upper hand** A cartoon section through the motor cortex of the human brain (*homunculus*) illustrates the great dexterity of the hand relative to other body parts. The area of the cortex which controls the motor functions of the hands is about as large as the area controlling the rest of the body from the shoulders down. After Penfield and Rasmussen, 1950, Fig. 22.

and their evolutionary implications (many of which have been already listed by John Napier, 1993). In the end, the overall dexterity of the human hand is best explained, of course, by the fact that it dominates the motor cortex of the human brain (see Figure 4.1). The extended fuctional plasticity afforded by the human hand would, indeed, be meaningless without the plasticity of the human brain that controls the hand's action. Taken together, these two sources of phenotypic plasticity entail an unprecedented degree of human diversity. The fact that humans also happen to deploy a system of division of labor and exchange unprecedented in scale and complexity is, of course, more than an evolutionary coincidence. Indeed, the pattern repeats itself, to a lesser degree, in some living beings other than human.

Next to humans, an organism that relies heavily on division of labor and exchange is the social insect. As expected, it displays a high degree of phenotypic plasticity (but no unusual degree of genetic diversity). Characteristically, individual members of a colony are divided into castes, each having a distinct function and markedly different body size and structure (see Figure 4.2). Within each caste, a given individual may perform different tasks during the course of its life, depending on age or environmental contingencies. Thus, it exhibits task-oriented plasticity

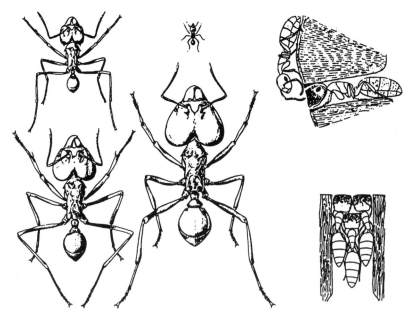

Figure 4.2 **Task-oriented morphological plasticity in the ants**
Specialized in tasks ranging from the tiny gardener to the giant soldier, the workers on the left are actually four adult sisters of leafcutter ants (*Atta laevigata*). The "living door" (top right) is about to step backwards allowing passage to her sister, a minor worker of the European ant *Camponotus truncatus*. Larger entrances (lower right) are blocked by several gatekeepers with similar door-shaped heads. Adapted from Oster and Wilson, 1978 (leafcutter), and from Szabó-Patay, 1928 (European ant), respectively.

not only in morphology but also in behavior. This pattern repeats itself in hundreds of species of wasps and bees, in addition to most species of ants and termites – not to mention certain species in the lower invertebrates that display even more striking examples of morphological plasticity oriented toward division of labor.

In the vertebrates, however, division of labor and, especially, task-oriented morphological plasticity are generally far less impressive. Group-living vertebrates occasionally, and to different degrees, may rely on division of labor in commonplace tasks: parenting, hunting, food transport, nest-building, guarding, and various roles of leadership. Task-oriented plasticity in these animals expresses itself primarily through sexual dimorphism but otherwise is largely confined to patterns of social behavior. Thus, for instance, patterns of dominance, territoriality, group size, mating, and parenting in certain species are occasionally altered in response to changes in predator pressure, population density, plant community, prey size, and other changes in food sources. While such reactive intraspecific variation is especially evident among the primate (Crook, 1970) and canid social systems (Moehlman, 1989), a list of several hundred examples in other vertebrates has been compiled by Dale Lott (1991).

Functional plasticity associated with morphological structures analogous to the human hand are not easy to come by in other mammals – especially, terrestrial mammals. The only example of such an exceptional structure of which I am aware is, perhaps, the elephant trunk. Elephants, of course, are highly sociable terrestrial mammals. Task-oriented plasticity in other species of mammals is largely confined, as indicated, to patterns of behavior. These are not easy to detect, given the free-spirited lifestyle of mammals. Yet, some indication of systematic task-oriented plasticity in mammals that rely on division of labor is provided by one of the earliest human experiments on animals: domestication.

The enigma within the enigma of domestication
Artificial selection served as a launching pad for the larger idea of natural selection. Titled "Variation Under Domestication," chapter 1 of *The Origin* opens with a seemingly mundane observation: "When we look to the individuals of the same variety or sub-variety of our older cultivated

plants and animals, one of the first points which strikes us, is, that they generally differ much more from each other, than do the individuals of any one species or a variety in a state of nature." There had been no plausible pre-Darwinian explanation for this problem of differential variation. But as Darwin and Wallace soon demonstrated, the enigma was readily reconcilable in light of the regularities of natural selection (e.g., Wallace, 1889:469). However, there was also an enigma within the enigma. Quite apart from the difference between domesticated and wild species there is the problem of differences among domesticated species themselves.

The ease with which breeding produces domestic varieties differs from one domesticated species to another, and so does the range of attributes over which breeding can systematically produce any discernable results. The dog, no doubt, is at the top of the list. Among hundreds of breeds, ranging in size from the Chihuahua (1.5 kg) to the Saint Bernard (100 kg), there is a bewildering variation in morphology (especially in the skull) and a systematic variation in behavior. The fact that in variations across breeds, dogs exceed by far – not only in degree but also in kind – any other species of domesticated animal is a complication that demanded attention. Could this added complication also be reconciled with the regularities of natural selection? "I do not believe," Darwin noted in a later edition of *The Origin*, "that the whole amount of difference between the several breeds of the dog has been produced under domestication; I believe that a small part of the difference is due to their being descended from distinct species." This view was later reiterated by Wallace (1889:88–89) and elsewhere supported with a detailed explanation by Darwin (1868:27–59). The argument from multiple origins was a risky proposition for both of them. The same argument was often used by detractors of natural selection attempting to cut off the theory from its main premise: the existence of intraspecific variations. The fact that Darwin and Wallace were willing to resort to this argument indicates, I think, the importance they both attached to the need to provide some explanation for a conspicuous case of seemingly excessive variation. Still, both insisted that the bulk of the variation observed across breeds of dogs is for the most part intraspecific.

Today, there is a wide consensus among experts that the dog is a descendant of a single species, probably the gray wolf (*Canis lupus*). Thus, on the larger issue of intraspecific variations, Darwin's and Wallace's common position was (somewhat ironically) vindicated. On the other hand, once the possibility of multiple origins was discredited, the elusive sources of variation that affect dogs has been left again without a plausible explanation. A number of salient observations about domestication are apt to help us address this question in the present framework.

> Observation no. 1: *On the whole, breeding operates most effectively on morphological and motor traits of animals (size, color, anatomical proportions, muscular power, speed, endurance, etc.). It operates less effectively on behavioral and temperamental traits.*

In fact, the relevant literature suggests that no new *behavior* pattern has ever appeared during domestication, and only in a few cases has a behavior pattern existing in the wild been lost (e.g., Immelmann, 1980:200). Patterns of behavior observed across different breeds of dogs can thus be treated – qualitatively if not quantitatively – as a fairly faithful description of their counterparts in the wild ancestor: the gray wolf. This point should be emphasized because, in the end, the main interest here rests with the products of natural selection rather than with the products of artificial selection as such.

> Observation no. 2: *All the large animals under domestication (above the size of the cat or rabbit) are essentially descendants of gregarious species, but only the dog is a descendant of a truly social species.*

Gregarious animals are more easily tamed, controlled and, above all, more easily amassed in a small space without resistance or ill effects. Thus, for obvious practical reasons, such animals are good candidates for domestication. What sets the dog apart, however, is that its ancestor – the wolf – is not merely a gregarious animal, but also an animal that relies on exchange and division of labor in the wild. The versatility of the dog under breeding should come, therefore, as no surprise. In light of the foregoing discussion, a relatively high degree of phenotypic plasticity is precisely what is expected of such a species.

> Observation no. 3: *Historically, the dog has earned not only the title "man's best friend" but also the well deserved title of "man's working partner."*

Going back at least to ancient Egypt in the forth millennium BC, dogs have been bred successfully for a wide range of special tasks. With little training or reinforcement and almost always with great enthusiasm, specialized breeds of dogs have served widely variant functions: hunting, guarding, herding, guiding, and drafting – to name but the most obvious categories. The main implication is that in large part the diversity observed in dogs is task-oriented plasticity.

> Observation no. 4: *The ease with which dogs can be bred and trained for a particular task, and the degree of motivation with which a task is performed, depend on whether the task has its direct or indirect origins in the lifestyle of the canids.*

Dog behavior is best understood in reference to its parallels in the wolf. These include, for instance, territoriality based on a recognized central headquarter (den) and its closely guarded surroundings; frequent excursions beyond this area; food storage, transport, and sharing; coordinated action under dominance order; and a long list of special skills geared towards cooperative hunting. Some insight into the latter was given by Darwin:

> . . . young pointers . . . will sometimes point and even back other dogs the very first time that they are taken out; retrieving is certainly in some degree inherited by retrievers; and the tendency to run round, instead of at, a flock of sheep, by shepherd-dogs. I cannot see that these actions, performed without experience by the young, and in nearly the same manner by each individual, performed with eager delight by each breed, and without the end being known – for the young pointer can no more know that he points to aid his master, than the white butterfly knows why she lays her eggs on the leaf of the cabbage – I cannot see that these actions differ essentially from true instincts. (1964:213)

A dog can be trained to perform tasks unrelated to its canid preadaptations. However, this can be achieved only with a considerable amount of conditioning and with constant need for inducement and reinforcement (in this respect, dogs do not differ from common draft animals or, for that matter, from laboratory or circus animals). The main implication, here, is that the task-oriented versatility observed in the domesticated dog is due in large part not to the ingenuity of artificial selection but to earlier structures produced by natural selection in a wild canid. The limits of breeding in producing truly new structures (unrelated to the tasks that already exist in the wild) is best demonstrated by the fact that despite the marked differences in behavior and morphology, there is little or no discernible difference in the level of general intelligence across various breeds of dogs.

> Observation no. 5: *Many breeds of dogs can be trained and motivated to act without the normal routine of food-reinforcement and physical punishment. Attention, praise, and petting are effective rewards, whereas expressions of disapproval and scolding are effective punishment.*[1]

This salient observation may help to provide, perhaps for the first time, a nonanecdotal and economically quantifiable indication for the existence of "emotional currency" in a nonhuman social mammal. The willingness of the dog to extend its services to humans at great expenditure of energy, effort, and inconvenience – sometimes at great risk to life – in exchange for nominal emotional rewards should not be confused with altruism or, alternatively, with some kind of a symbiotic adaptation. As typical of other mammals born in captivity, dogs identify their human handlers as parents or relatives of a sort, transferring to them all social relationships reserved only for close kin in the wild. Emotional currency has probably little value outside nepotistic markets. At a more fundamental level, this observation reflects the nexus between division of labor and exchange. It makes little sense for natural selection to produce adaptations geared to diversity and specialization unless it equipped the affected organisms with an adequate capacity to make exchanges. In the case of the dog, and the canids in general, this capacity is maintained by "emotional currency."

In sum, the dog provides an instructive case study that bears not only on a progenitor in the wild, but also on all social species. It unravels a tightly woven fabric of adaptations characteristic of a wolflike parent species into separate strands brought to the fore in various specialized breeds. It is true that some adaptations amplified in this process are specific to the canids. Others, however, are more generic in the sense of

---

[1] Such emotional rewards and punishments are not equally effective across all breeds of dogs. They are apparently less effective in certain breeds in the line of hunting dogs and more effective in other lines.

being applicable to almost any social species. The fact that a species that happens to rely on division of labor in the wild also displays a high degree of task-oriented diversity and a propensity to exchange cannot be dismissed simply as a canid peculiarity. The same configuration repeats itself in other social species. In addition to the wolf-dog we saw it in the social insects and, apparently by no coincidence, in humans. A three-way coincidence across such widely separated species is certainly worth thinking about. But coincidence or not, diversity and exchange is precisely what one expects selection to produce in animals that rely on division of labor.

The sexual division of labor

The changing roles played by gender cut not so much within as, on many occasions, across the two spheres of human subsistence: the domestic sphere of operation and the extradomestic (or market) sphere. For nearly 2 million years of hunting-gathering, it is now widely believed, the sexual division of labor was the central organizing principle of human procurement of food and other resources from nature. Hunting-gathering (and fishing) activities tend to favor at least partial separation of male and female day ranges. In most hunting and gathering societies, as far as we know, men and women are assigned to separate tasks of procurement – women are almost exclusively responsible for gathering and men for hunting and fishing (though in certain Arctic populations, men hunt whereas women fish).

Division of labor by gender quickly lost its cutting edge with the mass transition to farming and pastoralism some 10,000–6,000 years ago, depending on the region of the world. Tasks tend to be divided between men and women with far greater laxity in agriculture than in hunting-gathering. Physical demands still matter. For instance, weeding in subsistence agriculture is usually assigned to women whereas plowing that requires larger body size and greater strength is undertaken by men. However, activities are no longer rigidly compartmentalized into different spheres by gender. Martin and Voorhies (1975) who surveyed 104 horticultural societies and 93 agricultural ones found that women served as primary cultivators in 50% of the former but in only 15% of the latter. The implication is that gender roles are subject to considerable flexibility at least between the domestic and extradomestic spheres of economic activity.

Stringent sexual division of labor was resurrected, however, by the relatively cumbersome technology of steam powered machines introduced at the early stages of the industrial revolution, only to diminish again with the introduction of electric power and the advances in medicine. Electric power (and more recently hi-tech electronics) brought industry closer to home and enabled more flexible work schedules. Advances in medicine, especially in preventive medicine, prolonged considerably life expectancy. Taken together, these two developments helped to extend the effective working life span of women well into, and long past, childbearing age; so much so, that the nonagricultural labor force participation rate of married women (an inverse measure of the sexual division of labor) went up in most developed countries from about 5% early in the twentieth century to nearly 60% toward its end. All in all, sexual patterns of division of labor in human societies seem to be highly adaptable in response to technological and environmental changes.

The capacity to operate in grand-scale formations

The capacity to operate on a large scale is another adaptation that can greatly enhance the advantage of division of labor. Unfortunately, as we move to levels of organization at or above the unit of selection (typically, the "social" level), this capacity rapidly diminishes and in the vast majority of plants and animals it is practically nonexistent. Strange as it may sound, it is both an undeniable empirical fact and an inescapable inference from the principles of population genetics that the evolution of large-scale social structures – large-scale division of labor included – is primarily constrained by sexual reproduction. "Sex" points out E.O. Wilson "is an antisocial force in evolution" (1975). To be sure, the most imposing examples of division of labor in nature occur in settings quite innocent of sexual reproduction: in symbiotic relationships across widely separated (and obviously noninterbreeding) species, in colonies of invertebrates that reproduce asexually, and among the cells and organs of a single metazoan animal. With the fairly rare exception of monozygotic siblings (e.g., identical twins), no one under a system of sexual reproduction is perfectly related through common genetic descent to anyone, including parents and offspring. Conflicts in the distribution of (scarce) resources are almost inevitable. Sexual reproduction in this respect is an indirect hindrance to division of labor. It stands in the way of redistribution – the division of the *fruits* of labor rather than of labor itself – but the ultimate effect on division of labor is prohibitive all the same. A conflict-free (or conflict diffusing) built-in mechanism driven by the calculus of effort-rewarding transfers is a paramount adaptive prerequisite for any multiperson structure that exerts effort – if the structure is to be fixed in a population by selection. Division of labor is no exception. Without such a system of redistribution (through exchange or otherwise) division of labor would have little or no evolutionary implication, simply because its benefits could not be realized or properly internalized by the elementary units of selection: typically, individual members of a society.

Despite all hurdles, there are exceptions in nature – human affairs included – where large-scale division of labor operates under sexual reproduction at a fairly high degree of efficiency. These exceedingly rare instances are primary objects of curiosity, for each must harbor its own intriguing adaptive innovation.

Division of labor in insect society

Especially "innovative" in this class of adaptations is a toned down mode of sexual reproduction known as *haplodiploidy*. Among other things, it enables certain species of (hymenopterous) social insects – ants, bees, and wasps – to reproduce sexually, yet engage in division of labor and exchange on a large scale with little conflict. Under haplodiploidy, female offspring are conceived sexually, whereas males are derived from unfertilized eggs. Though males are reproduced asexually, their immediate parent (mother) and offspring (daughters) are always products of sexual reproduction. As such, haplodiploidy can hardly slow down the speed at which new combinations of mutually beneficial (or deleterious) mutations occur throughout a population. Haplodiploidy thus puts little or no drag on the rate of evolution – widely believed to be the main adaptive advantage afforded by sexuality – and at the same time helps considerably in the way of relaxing constrained social structures. For instance, it relaxes not only the constraint on large-scale division of labor but also the *sex-ratio constraint* (which is yet another example of a major adaptive disadvantage issued by sexuality at the level of groups).[2]

Haplodiploidy launches new configurations of kinship unknown under ordinary sexual reproduction. The calculus of relatedness under this mode of reproduction, as pioneered by Hamilton (1964 and elsewhere), offers highly parsimonious explanations to fairly awkward

---

Box 4.2 Sterile workers

Darwin, as well as R. A. Fisher, provided what now seem to be consistent, but only partial, explanations for the existence of sterile workers in the insects. Darwin viewed the question as one of the "difficulties" of his theory and proposed to resolve it by the fact that "selection may be applied to the family, as well as to the individual" (1964:237) – that is, essentially, by kin selection. Fisher took the argument a step further by making the, now common, analogy with the multicellular body: "The insect society more resembles a single animal body than a human society . . . the reproduction of the whole organism is confined to specialized reproductive tissue, whilst the remainder of the body . . . taking no part in reproduction" (1958:200). Both explanations are obviously consistent with inclusive fitness, as the concept was developed by Hamilton, but both also fail to specify the explicit adaptive mechanism that singles out the hymenopterous worker for infertility – that is, precisely, the mechanism driven by haplodiploidy.

Under haplodiploidy, sisters share a single diploid queen mother and, most likely, a single haploid father. Consequently, they are more closely related to each other than to their parents. Sterility is selected for, under these conditions, simply because it is genetically more expedient to help a mother bear a sister than to have a daughter of one's own.

---

[2] The constraint on sex-ratios, first formulated by Fisher (1958:158–160), implies that under ordinary selective pressures the two sexes will be produced in numbers close to parity. This principle of sex-ratio selection was modified and extended by Hamilton (1967) to account for many situations in which biased sex ratios can be expected, and are actually observed. For further extensions that bear upon group selection and the level-of-selection debate, see Wilson and Sober (1998).

questions (e.g., to the existence of sterile workers) that previously baffled the discourse of population genetics, going back to Darwin (Box 4.2). The fact that haplodiploidy causes sisters to be genetically more closely related to each other than to their parents sets in motion selective pressures that, on one level, favor sterility and, on another level, eliminate most (though not all) conflicts between siblings. It thus permits and, in fact, promotes expansion in the number of siblings living together (for the most part, sterile sisters) to the point where each colony is for all practical purposes, both, a single nuclear family and a population. Being confined only by considerations of economy of scale, the number of actors partaking in division of labor and exchange can swell with relatively little conflict – and great advantage – to any extent deemed optimal by selection. In short, haplodiploidy creates conditions conducive to division of labor on a grand scale.

The invisible hand

Mercantile exchange is an evolutionary innovation poles apart from haplodiploidy. But in a way, and by other means, it serves essentially the same function. As we have already had occasion to notice, exchange enables the division of labor in human society to operate on a large scale within and across populations. Here, division of labor no longer relies on the calculus of (genealogical) relatedness. It operates, instead, on a strictly private calculus that has some desirable, though (privately) unintended, collective consequences. It thus operates, so to speak, on the principle of *mutually beneficial selfishness*. Mutual benefits serve to defuse, or at least mitigate, otherwise prohibitive intraspecific conflicts. Selfishness may help establish the practice in an adaptive sense.

Through six editions of *The Theory of Moral Sentiments*, and through *The Wealth of Nations*, Adam Smith never in the least suspected his fellow human beings of being defective in self-love. I do not know if in his capacity as a moral philosopher he was overly inclined to worship human selfishness, but purely as an economist he certainly revered a *corollary* to selfishness: the paradoxical fact that action taken in the ordinary business of life is nearly as often benevolent by *implication* as it is selfish by *intention*. (More recently, economists such as F. A. Hayek and M. Friedman argued that this corollary works fairly well also in reverse.) Selfishness to an economist is very much what wind is to a sailor – neither a vice nor a virtue but a neutral force of nature – a force to be reckoned with for propulsion and, occasionally, for destruction. Properly rigged and harnessed by the institution of the market, the selfish motive power is an incitement to private action that benefits others as well as one's own self: "It is not from the benevolence of the butcher, the brewer, or the baker, that we expect our dinner, but from their regard to their own interest. We address ourselves, not to their humanity but to their self-love," as Adam Smith put it (1976:18). The conversion of sheer private selfishness (e.g., profit seeking) into socially desirable consequences (e.g., efficient allocation of scarce resources toward competing ends) is primarily due to a fairly obscure configuration of market forces – the *invisible hand of the market* – on which he is even more explicit:

> . . . every individual necessarily labours to render the annual revenue of the society as great as he can. He generally, indeed, neither intends to promote the public interest, nor knows how much he is promoting it . . . in such a manner as its produce may be of the greatest value, he intends only his own gain, and he is in this, as in many other cases, led by an *invisible hand* to promote an end which was no part of his intention. . . . By pursuing his own interest he frequently promotes that of the society more effectively than when he really intends to promote it. (1976:477, italics added)

What Adam Smith implies here is the existence of a self-regulating superstructure of invisible market forces that hovers over all bilateral market transactions (i.e., the only part of the economy that is actually under our direct awareness) and, since it operates entirely without guidance from an outside intelligence, it leaves occasional observers with little clue but to wonder about the origin of riches. For more than two centuries, since Smith wrote his piece, economists have been paying increasingly close attention to this invisible superstructure (or "hand"). By now we know that the formation of this seemingly mysterious structure is no miracle; though its workings are more complex and, in certain ways, even more wondrous, than Adam Smith envisioned. Every serious student of economics knows today to specify almost with mathematical precision the conditions under which the invisible hand is expected to function, or malfunction. (These conditions are, in fact, encapsulated in two proven mathematical theorems known in economics as the first and second *fundamental theorems of welfare economics*.) The silver lining – the plausibility of these conditions – can be easily grasped even without the aid of mathematics. It boils down, as I already indicated, to the simple fact that competition in the marketplace is essentially competition *for*, rather than *against*, trading partners. Producers compete for loyal customers of finished products and for reliable suppliers of raw and intermediate resources, professionals compete for clients of their specialized services, landlords likewise compete for tenants, employers for employees, and so on – the list is open ended and runs in both directions.

This should bring little comfort, of course, to hard-working entrepreneurs on the brink of financial ruin. More often than not, they tend to lose their businesses, and fortunes, to more seasoned rivals striving for the same customers. Though the overall business failure in the US economy since 1870 rarely exceeded the annual rate of 15 per 1,000 (Kurian, 1994), it was not unusual to see that on many occasions the appearance of a single innovative competitor – be it Henry Ford or Bill Gates – effectively drove the competition out of business by the dozens. However, neither Ford nor Gates could have possibly rated a mention had they chosen to promote their products by any means other than first, and foremost, promoting the interest of the final user – to the great benefit of millions. Indeed, the only proven way to win the competition in the marketplace is through the (selfish) practice of outcompeting the few by responding to the deepest needs of the many. One can blame Adam Smith for condoning selfishness under a veneer of economic respectability, but nobody can blame him for an inopportune choice of title for his major work: *An Inquiry into the Nature and Causes of the Wealth of Nations*.

The overall configuration of market forces that gives rise to the invisible hand has an evolutionary implication that thus far has escaped the attention of evolutionary theorists and economists alike. The fact that these forces regulate themselves through rewards and penalties (in the final stages of redistribution) exposes division of labor to selection pressures thus, in an adaptive sense, promoting it at a level of organization which is ordinarily shielded from such pressures. Indeed, the division of labor in a fine-tuned market economy is functionally always associated with an equally fine-tuned system of income distribution (with the understanding that, individually, members of a society ordinarily earn the value of their marginal productivity and pay marginal costs for the products each consumes, whereas collectively, the total value of the product is just exhausted by the total cost to produce it). As I have already indicated, the failure of division of labor in nonhuman societies to occur in any significant way above the level of the organism or its immediate kinship group is largely due to the lack of mechanisms, elsewhere in nature, whereby collective benefits can be properly internalized at the selection level of the constituents partaking in the larger enterprise. The invisible hand, as it turns out, provides such a mechanism. It operates smoothly at the level of markets – that is, at the level of populations and nations, and beyond – precisely because it issues benefits (properly calibrated) at the level of individual agents.

The striking analogy in division of labor (and other structures) that can be drawn between the (human) economic system and the workings of a body of a single multicellular organism bears testimony to the unique role of market forces in the course of human evolution. It is worthwhile noting, however, that the similarity goes nowhere beyond analogy. If we carry the division of labor in human society to its limit of refinement, we are not bound to end up with an organism of any form known in nature. The most apparent reason for this, as pointed out by Ronald Fisher (1958), is that human society, unlike insect society, contains nothing that nearly resembles or can possibly evolve into a specialized reproductive tissue. At a more fundamental level, it should be noted that the central organizing principle of the multicellular division of labor is the inhibition of selfishness (see Box 4.2). The division of labor in human society, quite the contrary, actually relies on selfishness and (in an evolutionary sense) promotes it. So, we end up with an organism as yet unknown.

5　The feeding ecology

The incredible shrinking gut

Laboratories under sail

Among the sailors who crisscrossed the oceans during the sixteenth century few knew how to swim. It was pointless. A sailboat at sea could not be stopped or turned to rescue a shipmate washed or fallen overboard. The hazards of the occupation could not escape the attention of sailors struggling to lash unruly sails in unpredictable gale-force winds. One slip on the upper yards meant almost certain death. Given the state of the art and the perils of the sea, casualties from accidents and disease were exceedingly high even by the standards of the time. As to comforts and niceties of life – nobody could ask for, nor possibly be granted, accommodations beyond the bare minimum.

With lives aboard more expendable than cargo, one can wonder why skippers and ship stewards paid meticulous attention to supplies of foodstuff. A typical list of ship supplies for a journey across the Atlantic around 1577, the year Sir Francis Drake set out to circumnavigate the world, included such diverse items as hardtack, flour, pickled beef and pork, dried peas and codfish, butter, cheese, oatmeal, rice, honey, and vinegar, as well as about eight tons of beer.[1] The selection seems to compare quite favorably even today with any decent hospital or school cafeteria – not counting the beer. Careful attention to an adequate *bulk* of food supplies before embarking on unpredictably long trips such as the one undertaken by the *Mayflower* (66 days across the Atlantic in 1620) is perfectly understandable. But why overburden the already hard-pressed logistics of a ship on high seas with a wide *variety* of food items? Pure culinary considerations of fine catering could not possibly have been a

---

[1] There was, apparently, a practical reason for the relatively large supplies of alcoholic beverages carried by oceangoing ships of the time, and routinely consumed at alarming rates by the crew (once in a while at the cost of falling overboard even beyond the call of duty). The need to furnish sufficient drinking water on long trips was a major logistic problem, for fresh water could not be kept potable for long. Alcoholic beverages (typically, cheap wine and rum, in addition to beer) seemed to travel long distances much better than water.

high priority on a ship carrying one crewman for every 10–15 ton of cargo (compared with one for every 3,000 or more today). Instead, the need for diverse provisions was more nearly rooted in practical experience. Hard and not always pleasant were the lessons learned by mariners over 2 millennia of plying trade routes between coastal ports. But the main lesson was learned in the first century of extensive cross-oceanic sea exploration when voyages could go on for months at a time without fresh supplies.

Virtually a world on its own, a voyage on the high seas was a perfectly isolated feeding ecology. In a matter of a few weeks after setting sail, each ship became a floating laboratory unwittingly experimenting in digestive physiology (the human subjects rarely enjoyed the experiment and quite often did not survive). For instance, embarking into the yet unnamed and vastly underestimated Pacific Ocean in 1520, Magellan and his expedition crossed open waters for 3 months and 20 days without getting any kind of fresh supplies. Eventually they succumbed to eating ox hides and sawdust, and also hunted for rats on lower decks. However, the 19 shipmates who perished during this leg of the voyage were all, apparently, the victims of scurvy or some other nutritional deficiency, rather than starvation. The cost of sea exploration in terms of human life was high, but lessons were learned. The big lesson – that the Earth is round (for which a longsought practical proof was finally given) – was accompanied by a small sobering lesson that nobody sought to learn; namely, that in a matter of few weeks, the key to human health and sheer existence is diversity in food intake. Bulk does matter, but so does the mix. It is the smaller of these two lessons which matters most to our attempt to fully understand the origins of hunting-gathering as well as the challenges faced by the people who first made the transition to agriculture and food production.

The lesson learned on the waters was never fully appreciated on land. It took the British Royal Navy nearly 50 years to add (in 1795) lime juice to the rations of sailors following repeated recommendations by the Scottish naval surgeon James Lind. Lind, who was first to advance the concept of deficiency diseases, found this simple but effective preventive measure and cure for scurvy as early as 1747. Though the most conspicuous, scurvy is only one of a long list of nutritional deficiency disorders associated with lack of diversity in food consumption.

The economic approach to food consumption

Economics was even slower than the Royal Navy. Diversification in consumption, especially in food consumption, became essential to the way economists think about the formation of human preferences only during the last quarter of the nineteenth century. There were earlier precursors, but the discourse reached a critical mass only when the concept of *diminishing marginal utility* was independently and almost simultaneously made a primary target of inquiry by three of the best-known economists in the generation immediately following John Stuart Mill: Léon Walras (1834–1910) in France, William Stanley Jevons (1835–82) in Britain, and Carl Menger (1840–1921) in Vienna. Diminishing marginal utility is essentially the idea that the utility of an incremental quantity diminishes with an increase in the total stock. As such, its main implication suggests eclectic behavior on the part of consumers: the tendency to spread expenditure over a large number of commodities (especially food items), consuming each in small quantity, rather than large quantities of only one or a few. Contemplating the alternative (i.e., constant or increasing marginal utility) implies consumption of a single commodity. With the possible exception of acute cases of addictive behavior, such an extreme form of specialization in consumption is rarely observed in humans, although it is not unusual in other animals (e.g., the koala which feeds only on the leaves of the eucalyptus). The concept of diminishing marginal utility was greatly refined in the first part of the twentieth century (and replaced with the concept of *diminishing marginal rate of substitution*) though the major thrust remained closely associated with diversification in consumption. Economists today rarely deal with the consumption of a single commodity in isolation. Instead, they are inclined to think about consumption in terms of "baskets": bundles of goods and services consumed in combination. Diversity, as a matter of empirical reality, seems to be the central organizing principle of human consumption. Not independently, diversity is also essential to the way economists think and reason about this reality. Although economic theory does not single out consumption of food for special treatment, it is clear that this is where the lesson of diversity applies with far greater force than in any other form of consumption under its investigation.

An adaptive approach to food consumption

By its very nature, food consumption falls in an area where behavior and morphology closely overlap. Readily observed parallels between feeding behaviors on the one hand, and the corresponding anatomy and physiology of digestive systems, on the other, lend themselves to *testable* adaptive explanations. There is also a growing agreement among experts about the existence of a strong (inverse) relationship between expanding brain tissues and shrinking gut tissues (in the evolutionary trend of encephalization), which makes the entire analysis all the more intriguing from the viewpoint of human evolution (Aiello and Wheeler, 1995).

By descent, humans like most primates are omnivores. Unlike a carnivore that operates only on the upper trophic level of the food chain (eating primarily animal matter), or an herbivore that operates only on the second trophic level (eating primarily plant matter), an omnivore operates on both levels (eating both kinds of food). The differences among these three strategies in the feeding ecology of *mammals* are more fundamental and encompass more than merely the content of the diet. The advantages and disadvantages of a carnivorous diet compared with a herbivorous one can illustrate the point.

The main disadvantage of being a carnivore, like a tiger or a weasel, is that the food is mobile – it runs away and, occasionally, it even fights back. The acquisition of animal food entails large expenditures of energy and expensive hunting adaptations. Once acquired, however, the meals are readily digestible and nutritionally balanced. By comparison, herbivores whose diet consists mainly of sessile and abundant plant matter spend relatively little energy on procurement. The main challenge faced by avid vegetarians like horses or antelopes – or even gorillas – is in digesting and detoxifying their food. Leaves, for instance, are the most abundant and easily accessible potential source of food (including protein) to be found on dry land, but also the least digestible. Leaves are often laced with digestive inhibitors and toxins and contain cellulose and other structural carbohydrates that can be broken down only by bacteria housed in the gut (and require special and fairly expensive anatomical and physiological adaptations on the part of the host). Moving down from the sanguine top to the verdant lower levels of the food chain means, therefore, great savings in the cost of procurement at the expense of greater costs of digestion and processing. The tradeoff is evident in the size and proportions of the digestive tract. Smaller and more simple, the guts of carnivores are proportionally dominated by the small intestines (emphasizing absorption). The guts of herbivores, leaf-eaters in particular, are more complex and generally are dominated by a proportionally enlarged stomach or caecum and colon (emphasizing fermentation). Most complex, of course, is the gastrointestinal tract of ruminants (deer, cattle, sheep and goats) with its manifold of elaborate fermenting pouches and chambers. Where do humans fit in this picture?

The shrinking human gut
Judging by the content of the diet, humans are omnivores; to be sure, we eat both animal food and plant food. But when we try to place ourselves between opposite ends of the continuum from meat-eaters to plant-eaters, we soon reach a paradoxical observation. Very few humans take the bulk of their diet from animal matter except, perhaps, in high arctic latitudes where the scanty vegetation has left the local (Inuit) people with little choice but to acquire special adaptations to diets high in fat and protein derived from animals – sea mammals and, to a lesser extent, fish and caribou – the only food sources that, historically, were available to them. World wide, however, meat usually constitutes a very small proportion of the total human intake of food (Harris, 1992). Most humans take the bulk (wet weight) of their diet from plants and complement it by a steady but modest amount of animal matter. Judging from what is known of the diets of recent hunter-gatherer societies (Lee, 1968) and from analyses of human coprolites (Kliks, 1978), as well as from the analogy with the diets of extant primates, one can safely surmise that meat-less meals were the rule rather than the exception in past evolutionary times and, especially, in early agricultural times. It would be reasonable therefore to anticipate that the human digestive system will come fairly close to the model expected of an herbivore or, at least, of an herbivorous omnivore such as the chimpanzee. However, this theoretical expectation falls apart as soon as we are confronted with data about the actual structure and dimensions of the human gut.

The human gut is markedly small relative to body size and relative to similar metabolically expensive organs in the human body: the heart, liver, kidneys, and lungs – not to mention the brain. In fact, it has been estimated (Aiello and Wheeler, 1995) that the total mass of the human

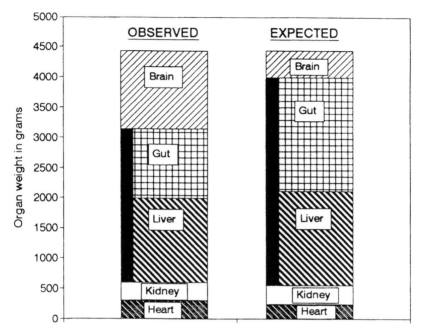

Figure 5.1 **Expected and observed size of metabolically expensive tissues in the human body** Note the remarkable tradeoff in relative size between the brain and the gut. In contrast, the size of the heart, kidneys, and liver are hardly affected. Adapted from Aiello and Wheeler, 1995.

gastrointestinal tract is only about 60% of that expected for a similar-sized primate (see Figure 5.1). Moreover, almost 2/3 of the (already reduced) volume in the human gut is taken up by the small intestine, compared with less than 1/3 for the great apes in general (and less than 1/6 in the gorilla). In other words, as Chivers (1992) pointed out, the human gut dimensions are those of a faunivore (i.e., a meat-eater). This raises the serious question of compatibility of an organ with its primary function.

Externalization of function
The enigmatic shortfall in the human digestive system can be explained in part as an incidental evolutionary effect of *externalization of function*. Manual processing (removing dirt, peeling, husking, cracking, crushing, etc.) and chemical modification (mainly through cookery) of dietary

items before ingestion effectively buffers the digestive system against mechanical stress, excess bulk, and toxins. The dexterity of the human hand in combination with tools and, above all, in combination with the control of fire, had in this respect a profound effect on the digestibility of human diets, and by extension, a corresponding economizing effect on the human gut and on human dentition.

The "skin of its teeth" is an animal's first contact with its ingested food. The fact that the enamel coating of teeth is the hardest biological substance known bears testimony to the mechanical strain in this stage of digestion. It can thus be argued that the externalization effect would ease, first and foremost, the strain in the mastication stage of digestion. There is indeed a consistent trend of reduction in dentition, species to species, up the evolutionary line of the genus *Homo*. However, by comparison, the differences in dentition between some of the Australopithecine species (e.g., *A. africanus*) and modern humans seem not far greater than differences within the *Australopithecus* lineage itself (e.g., between *A. afarensis* and *A. boisei*). Using the extant chimpanzee as benchmark, the differences in the chewing apparatuses appear in some respects to be less clear-cut than in other parts of the digestive system. Of course, the differences in dentition between humans and chimpanzees are great – far greater, in fact, than those between the chimpanzees and the gorilla. However, these differences are not necessarily traceable, in any systematic way, to the externalization effect. Reduced incisors and canines in humans, but enlarged cheek teeth (premolars and molars), seem to cut the issue both ways. Least compelling in this respect is the enamel coating which, contrary to what might be expected, is far thicker in humans than in the chimpanzees. The weight of these observations suggests the possible existence of adaptive pressures operating primarily on the lower (gastrointestinal) parts of the human digestive tract that have little to do with the oral cavity; that is, forces at work in addition to, and quite apart from, the incidental effect of *externalization*.

The expensive-tissue hypothesis

No longer incidental is an adaptive effect recently suggested under the *expensive-tissue hypothesis* (Aiello and Wheeler, 1995): the idea that only animals with cheap guts can afford expensive (i.e., large) brains. It suggests that the metabolic requirements of a relatively large brain must be offset by a corresponding reduction of the gut. The enlarged human brain is a highly expensive organ in terms of the quantity of energy it consumes. It costs us roughly 16% of our total basal metabolic rate (BMR).[2] To fuel (with substrate and oxygen) such a large brain and still maintain a normal BMR, another expensive metabolic tissue must be reduced. The key argument is that this has to be the gut.

In fact, the increase in mass of the human brain appears to be balanced by an almost identical reduction in the size of the gastrointestinal tract. Now at 15% of total BMR (reduced, presumably, from a previous total of 25%), the gut is the only body part that could have possibly varied in size sufficiently to offset the metabolic cost of the encephalizing human brain. The reason for this is that other metabolically expensive organs – liver, heart, kidneys, and lungs (accounting for 19, 11, 8, and 4% of total body BMR, respectively) – could not be reduced since they are physiologically constrained in fairly strict proportion to body size. Gut size is only partly related to body size. Its size and proportions (in a species) are in large measure also determined by diet, that is, by behavior. It follows that the relatively large brains in humans (and to a lesser extent in other primates) could not have been achieved without a shift to a high-quality diet. In this respect, Aiello and Wheeler emphasize the consumption of greater quantities of meat relative to plant foods as a likely possibility.

Consumption of meat: pro and con
Viewed strictly from the standpoint of digestibility, there are certain advantages to the consumption of meat and animal matter in general. The fact that the digestive tract is relatively smaller and far more simple in carnivores than in herbivores bears testimony to this effect. Meat contains *high-quality* proteins that provide all the amino acids in the proportions required for the human body. These include nine essential amino acids that can not be synthesized internally. Proteins derived from plants lack one or more of these essential amino acids and in this sense are considered as *low-quality* proteins. In addition to proteins, meat provides in

---

[2] Estimated shares of BMR consumption in the present discussion are rounded to the nearest percentage point (based on data presented by Aiello and Wheeler (1995, Table 1) and originally adapted from Aschoff, Günther, and Kramer (1971)).

the same concentrated highly digestible capsule almost all the other nutrients essential to mammalian diets – carbohydrates, lipids, minerals, and vitamins. However, the benefits of flesh-eating can be fully realized only by true carnivores that possess special physiological adaptations. Tigers have little to worry about regarding cholesterol, and few eagles, apparently, die of clogged arteries. But humans do.

With the (intestinal) anatomy of a carnivore but the physiology of an omnivore, the benefits of flesh-eating that occur to humans are less clear-cut. Consumption of small or moderate amounts of protein-rich animal foods is undoubtedly beneficial (to the eater). It fulfills the daily amino acid requirements, alleviating the burden on the digestive tract of processing large quantities of poorly digestible plant matter, typically, proteinaceous leaves. As such, it helps a smaller gut mass support a larger body mass. However, protein-rich animal foods are less desirable as a source of energy, for which there are superior substitutes in the form of highly digestible carbohydrates derived from plant matter, typically, fruits and seeds. To meet the human energy requirements exclusively from meat, especially from lean meat, an eater would have to consume exceedingly large amounts. But for most modern humans, large quantities of animal protein may actually be detrimental to both normal growth and good health (Milton, 1995; Edozien and Switzer, 1978; Nelson, 1975). In other words, the marginal benefit from the last unit of animal protein consumed is initially fairly high, but then rapidly diminishes as additional units are added to the diet and, eventually, it becomes negative. It follows, therefore, that beyond a certain point meat rapidly exhausts its usefulness in human consumption. It is not by coincidence that even when given the choice (e.g., in affluent societies) most people still prefer to consume meat in amounts that rarely exceed 40% of the total food intake.

In the actual physiological and ecological context of feeding there are certain complicating factors that economists are usually unaware off, although such complicating factors may have non-trivial bearings on their models. One complicating factor associated with the human physiology of digestion is its tendency to operate in a hierarchical (or lexicographic) order of priorities – starting with energy before it gets to the business of proteins. Served to a hungry person, lean meat (or skim milk) is bound to be converted to calories in short order. Proteins would be processed for body building and repair only when the energy deficit has been fully restored. To conserve scarce proteins for their intended function in the human diet (the provision of amino acids, nitrogen, and certain vitamins which cannot be obtained from plant foods), animal foods are best consumed in combination with ample amounts of either fat or carbohydrate and ingested in an appropriate temporal order in the course of a day or even a single meal. There is no need for conscious planning on the part of eaters. The configuration deemed optimal by a long evolutionary experience on the brink of starvation is already hard wired in taste buds and cravings. Driven by adaptive tastes, few people choose to consume meat (especially when scarce) as the first meal of the day, or the first course of dinner; and most prefer to eat spaghetti and meat balls together rather than to feed on the two ingredients separately. In that, they certainly obey an unwritten universal law: the dietary law of conservation of proteins. In this respect the marginal benefit from the last unit of meat consumed is a function, as economists are first to recognize, not only of the total quantity but also of the quantities of all other foods consumed in combination. In addition, the marginal benefit is also dependent on the order and timing in which the different items are consumed. The dynamics of ingestion are by no means cost free.

The transition to hunting-gathering
Our Forebears, at some sufficiently remote point in natural history, almost certainly relied on the feed-as-you-go mode of subsistence. (This mode of subsistence, in all likelihood, was still in place throughout much of the Australopithecine stage, roughly 5–2 million years ago.) Under a feed-as-you-go strategy of feeding, as typical of most primates and many other foraging and grazing animals, food items are ingested on the spot and strictly in the order of acquisition. Under this strategy, the route a feeder chooses to follow in its daily ranging largely determines its meals for that day. If the route is to minimize the energy cost in procurement, it will minimize the energy costs of digestion only by coincidence. Conversely, if the route is to minimize the cost of digestion, then the feeder runs a serious risk of starvation simply because the expenditure of energy and time in procurement are bound to be arbitrarily large. The best route (and thus diet) under the circumstances is a compromise that minimizes costs neither in procurement nor in

digestion. The good news about this situation is that it leaves room for evolutionary innovations.

The most obvious innovation, we now understand (with some benefit of hindsight), was to replace the linear feed-as-you-go strategy with the strategy of a hub. The hunting-gathering-fishing-etc. hub is the idea of separating in space and in time the act of procurement from the act of ingestion and thereby, simultaneously, minimizing the cost of both. Minimizing the cost of procurement is essentially achieved through division of labor and specialization in ranging. Minimizing the cost of digestion is achieved by increasing diversity of food items, improving the mix of the diet and the timing of ingestion. Radial ranging, consumption postponement, food transport to a central place, as well as food-sharing of sorts – i.e., almost all the elements of such a hub – have been inferred from the Oldowan data dating nearly 2 million years ago, as presented and interpreted by Glynn Isaac and his colleagues (to be discussed in Chapter 8).

It should be noted that division of labor and specialization in the procurement of food and, especially, the flexibility in ranging granted to hunters, was perhaps the best possible mechanism that could economically facilitate an increase in the provision of meat and, consequently, an improvement in the quality of the diet. Whether such an increase actually took place (either in specific local settings, or in the general evolutionary trend) is, of course, an empirical question. The sum of the findings generally seems to suggest that this was the case.[3] However, the overall improvement in the quality of the diet associated with the departure from the feed-as-you-go practice of feeding was probably far deeper and more profound. The increase in the consumption of meat, in the final analysis, was only one part of it. This view falls not far from the position taken by some experts in the fields of dietary ecology. For instance, Katharine Milton, a leading authority on the digestive physiology of primates and a source from which I have benefited throughout the present discussion, makes the following points:

---

[3] The findings in support of this conclusion come from diverse sources: e.g., the reduction in human dentition, the large assemblages of animal bones in juxtaposition to manufactured stone tools, the analogy with the diets of chimpanzees and savanna baboons, the major theme in cave art (i.e., grazing animals), and more.

> Like some other researchers . . . I see a division of labor with
> respect to food procurement in combination with food sharing as
> a pivotal adaptation in human evolution. Indeed, I think that the
> implications of this type of dietary innovation have not been fully
> appreciated, for, in effect, a division of labor and food sharing
> provide a means whereby individuals of a given species can
> efficiently utilize foods from two trophic levels simultaneously
> – a foraging strategy that appears to be truly unique among
> mammals. (1987:108)

Indeed, it can be argued that the spectacular replacement of gut tissue
with brain tissue (in combination with an overall increase in body size
and no loss in metabolism) seems to imply a more dramatic change in
dietary strategy than merely a modest, and highly opportunistic,
improvement in the intake of meat. For all its expediency, the growing
reliance on meat can be viewed as only one element in the broader and
more comprehensive breakthrough in the dietary strategy.

In sum, the departure from the feed-as-you-go strategy to hunting-
gathering was associated, it seems, not only with a quantitative improvement in diet but also with an improvement in its quality – especially, in
the intake mix and in the timing of ingestion – and thus, quite possibly,
with a great economizing effect on the digestive system. Consequently,
under the Expensive-tissue Hypothesis, it is safe to assume that the transition to the hunting-gathering feeding ecology facilitated a growing
brain (and, no doubt, was facilitated by it). Empirically, the timing could
not have been more supportive of this conclusion. Long before the
Expensive-tissue Hypothesis was brought to our attention, the anthropological and the paleoarchaeological records already traced the onset of
encephalization and the approach of hunting-gatherings to the same
period in time (roughly, 2–1.8 million years ago), to the same place (in
Eastern Africa), and to the same ancestor (*Homo habilis*).

Life-cycle versus evolutionary consequences

My discussion concerning food consumption was driven by implications
on the time scale of evolution, as distinct from life-cycle implications on
the scale of behavior. The bearings upon current diets and dietary issues
brought to the fore by certain modern lifestyles (e.g., obesity, substance
abuse, anorexia) are tangential and may even seem to be immaterial. The

reason for this is that the digestive system possesses wide margins of redundancy, as is probably true of other structures in the human body. Adult males and (nonpregnant) females may endure for weeks, if not for years, on a suboptimal diet without showing much in the way of long-term ill effects. Adaptations matter most, and thus tend to be established, when margins of redundancy are exhausted to the brink of survival, typically, in situations that require the utmost exertion of physical effort: running for life, fighting disease, as well as enduring dehydration, hypo- and hyperthermia, and all kinds of trauma. Margins of redundancy are far narrower in the case of children and adolescents. Malnutrition, in this case, has enduring ill effects on growth and development with obvious implications not only in evolutionary trends, but also in the life cycle of individuals in a single generation. More subtle, by definition, are certain dietary problems that pass unnoticed in the life cycle of an individual, but may have consequences of great import in the evolutionary process of the species at large. For instance, certain deficiencies (e.g., of essential fatty acids) in the diet of women may lead to irregularities in the menstrual and reproductive cycles. Trivial episodes in the theater of a lifetime may have major repercussions in the theater of evolutionary events (and vice versa).

Runaway arms races in a vertical feeding ecology

The brain consumes eight times its share in metabolic energy, at rest. Indeed, as we saw, human beings expend about 16% of their basal metabolic energy to fuel a brain that weighs an average of 1.3 kg (roughly 2% of body weight). At this size, the human brain is already twice as large, and twice as expensive, as that of a primate similar in size. There must have been a compelling adaptive reason for such an expensive tissue to double in size over a relatively short evolutionary time span (i.e., 2 million years, approximately). I have already mentioned (in Chapter 3) the possibility that exchange was closely involved in this costly process of brain expansion. The mechanism suggested is a self-reinforcing process (in the form of an arms race) that operates in the arena of human subsistence, for that is where exchange takes place. The process is better known, and more vividly observed, in the arena of sexual selection where it tends to produce exaggerated body parts such as the celebrated tail of the peacock and (seemingly) suboptimal behavior, typically in males. Under sexual selection, the process is primarily driven by male to male competition for mating partners. There are, however, instances where such a self-reinforcing process takes place in the course of competition for food, rather than for mates. The two examples I will discuss below should be evaluated as analogies to encephalization, except that the enlargement is in height rather than in brain volume. At issue is the evolution of the tallest living animal and the tallest living plant: the giraffe and the *Sequoia*, respectively.

The giraffe
Reaching some 6 m (20 ft) above the ground, the giraffe is a large browser that requires up to 66 kg (145 lb) wet weight of ingested foliage a day to sustain its bulk (males exceed a metric ton). This amount of vegetation is not always easy to come by in the giraffe's range – the northern and southern arid zones of the African savanna. In dry seasons and during sporadic droughts, giraffes, like many other savanna herbivores, find themselves in head-to-head competition for the dwindling vegetation along the few remaining and increasingly overcrowded watercourses. Hard times call for special adaptations. Elongated forelegs and neck, a vertically tilting head, and a far-reaching tongue (a grasping organ as long, and nearly as versatile, as a human forearm and hand) make up a set of adaptations that keep at neck's length all unwelcome competition from below. Indeed, with the ability to feed off treetops giraffes face little competition from other browsers, let alone grazing animals. It is tempting to assume that the giraffe acquired its impressive stature in the course of a fierce struggle with other species for foliage; that is, through *interspecific* competition. But interspecific competition is not nearly as strong a force in evolution as some old textbooks (using the neck of the giraffe as an example) seem to imply.

The *Hawk–Dove Game* is used (in more recent textbooks) to introduce basic ideas of evolutionary game theory. Although, strictly speaking, hawks and doves belong to different species, I am aware of no application where the game is played across species in an interesting manner.[4] In its most typical renditions it is actually intended to apply to

---

[4] Oddly enough, in one of its versions, the "Hawk–Dove" game boils down to what is known in economics as the game of "Chicken" – yet another species.

intraspecific, rather than interspecific, conflicts. Intraspecific competition is certainly the rule, rather than the exception, in the feeding ecology of almost all animals and plants. This rule gains added force in the elevated feeding ecology of the giraffes where the struggle between individuals is second only to their lifelong struggle with gravity. To pump up enough blood to the brain through a carotid artery 3 m (10 ft) long, the giraffe needs an expanded heart capable of maintaining blood pressure twice to three times that of a human being. (A metabolically expensive tissue to begin with, the giraffe's heart weighs 11 kg (25 lb), is 60 cm (2 ft) long, and has walls up to 8 cm (3 inches) thick.) The size of the heart relative to body bulk and in proportion to other metabolically expensive organs is perhaps the real measure of evolutionary cost incurred by the giraffe's lavish anatomy. One need not be a bioengineer to wonder if the energetic and structural costs of this anatomy are fully justified by the benefits.

The tallest basketball player standing on the shoulders of the tallest elephant could hardly reach as high as an average (male) giraffe. Indeed, the giraffe is not only the tallest living animal, it is also taller than the next tallest by a margin that defies continuity (6 m compared with 4 m of the African elephant). This gives the giraffe free access to a hanging bed of foliage nearly 2 m beyond reach of all other land browsers – an impressive advantage by any ecological standard; except that it seems to have partly exhausted its usefulness. Specifically, it leaves one-third of the giraffe's height (and two-thirds of its carotid artery) unexplained by competition with other species. The implication is that the expensive structures of the giraffe have long outreached their net survival value to the species as a group, though not necessarily to the individual. The only compelling reason these structures could actually work in favor of individual fitness (as distinct from group fitness) makes sense, it seems, only under a scenario of *intraspecific* competition analogous to a runaway arms race: an escalating process impelled by a self-reinforcing mechanism.

The mechanism is best known, as I have already indicated, from the arena of sexual selection (Maynard Smith, 1978), but it is by no means exclusive to it (Dawkins and Krebs, 1979). Sexual selection produces vivid examples of dimorphism, not least the tail of the peacock male. Such structures often seem to fly in the face of natural selection – assuming one chooses to view them through the prism of group selection to the exclusion of individual selection (or, alternatively, through the prism of phenotypic survival to the exclusion of genotypic fitness – a transgression not uncommon in classical Darwinism). In addition, and quite apart from sexual selection, the sexual division of labor operates in an adaptive sense to produce its own manifestations of dimorphism by gender. Which of these forces is primarily responsible for the extravagant neck of the giraffe, especially that of the male giraffe?

The fact that male giraffes can actually reach nearly a meter higher than females extends considerably the feeding range of bulls compared with cows. With such (potentially) far-reaching consequences to subsistence, it is hard to see how this gap in stature could have been produced simply by sexual selection. Sexual selection may lead to dimorphism that takes many forms, but *systematic* bias (against females) in the aptitude of reaching scarce sources of food is not one of them. If anything, because of higher energy cost (per unit of body weight) incurred in pregnancy, lactation, and parental care in general, females may be expected to be more dependent on such sources; thus, even more capable and efficient in reaching them. Indeed, they are. As a proportion of body weight, females are estimated to have a greater relative rate of food (dry weight) intake than males: 2.1% per day for cows compared with 1.6% for bulls (Macdonald, 1999:536). The discrepancy should not be taken as an indication of failure on the part of males in taking full advantage of extra feeding opportunities. Neither neglect nor self-deprivation can be blamed on an animal that actively feeds almost half of its time round-the-clock (43% compared with 55% for females, including nocturnal feedings under moonlight) and spends much of the remaining time ruminating. The discrepancy can be traced, instead, to a more fundamental adaptive principle – sexual separation in ranging – which in the final analysis also provides a compelling explanation for the special kind of dimorphism displayed by giraffes.

Sexual separation in ranging is an adaptive arrangement actually observed in many large animals that rely on sparsely distributed sources of food. Its main function is to reduce male (feeding) competition with their own offspring and mates through the use of separate feeding sites. Giraffes are an exception (among such terrestrial animals) only in that they adapted this universal principle in vertical dimension.

To be sure, giraffes of different age and sex feed off trees at different heights. For instance, it has been pointed out (Pellew, 1984; Macdonald, 1999) that when giraffes are viewed even from some distance, one can still tell them apart by gender simply by observing their feeding posture: bulls tend to browse high while females bend to take advantage of regenerating branches below. Moreover, growing tall offered the giraffes an opportunity to add a little unexpected twist to this arrangement between the sexes; their very own form of pseudo-agriculture. By munching on topmost leaves male giraffes in effect constantly prune their trees (mainly acacia) from above; thus, keeping the trees short and bolstering lower growth to the great benefit of females and calves.

Bulls tend to minimize feeding competition not only with mates and offspring, but also with parents and sibs. Thus, for instance, they migrate from their natal group at puberty (typically, at age 3–4), long before reaching full maturity (at age 7–8). Bulls occasionally also feed in heavily wooded areas that, for reasons of predation, are inaccessible to cows and calves. All these arrangements of separation in ranging get greatly complicated, however, by the fact that breeding in the giraffes is perennial. Cows conceive throughout the year (though more often during the rainy seasons). Males, by implication, must constantly monitor the reproductive status of females in their range. From the point of view of the male's reproductive strategy it is essential to identify females in estrous as soon as possible and keep (subordinate) rivals away. Selection pressures favor, therefore, male and female feeding ranges that closely overlap horizontally but, not necessarily, in the vertical dimension. That is why a bull will spend much of its time patrolling cows' ranging sites but, while there, will feed aloof beyond their and their calves' reach.

Putting all these pieces together, it is clear that the reproductive success of a male giraffe heavily relies on two partly conflicting principles. To minimize competition for scarce sources of food, the first principle is to keep maximum distance from close relatives; not the least from mate and offspring. The second principle, as we saw, calls for maximum time in attendance and close "proximity" (when giraffes mingle they rarely get closer than 20 m to each other and, at other times, 1 km is close enough). These two conflicting principles can be reconciled, it seems, only by vertical separation between bulls and cows; that is, by males growing taller – and this is taken care of by selection. Other things being equal, a slightly taller male can stay slightly closer to females (i.e., avoid competition with mates and offspring without loss of consort opportunity); namely, it is in a slightly better position to leave more offspring in the next generation. No male under such conditions can afford – in an evolutionary sense – to stay even slightly shorter than others. All the elements of a self-reinforcing mechanism that could, and probably did, propel the vertical arms race among the giraffes are now in place. It differs from similar processes propelled by sexual selection in a number of ways. Chief among them is the fact that it has actually produced a form of dimorphism that is largely instrumental rather than deleterious (although strictly from the *phenotypic* viewpoint of male survival it has probably long reached suboptimal proportions). Why did such a self-reinforcing process come (some 2 million years ago according to the fossil record) to a dead stop?

All arms races in nature eventually subside and get settled, either in a new state of suboptimal equilibrium or, perhaps, in extinction. By its very nature, any self-reinforcing mechanism that progressively consumes scarce resources must come equipped with its own self-arrest (or self-destruct) device. In the case of the giraffe this device is fairly transparent, as we can infer from the fact that with a linear increase in body stature, body bulk – and hence, food consumption to fuel it – must increase exponentially. Up to a critical point, an extra inch in the stature of bulls would mean net savings that leave extra foliage at the disposal of cows and calves. Beyond this point, any additional inch in height would actually entail only net losses. In the habitat of the giraffes this critical point was apparently reached 6 m above the ground, and 2 million years back in time.

The sequoia tree

The excessive height of sequoia trees, especially the coast redwood of California (*Sequoia sempervirens*), has probably evolved by much the same self-reinforcing process: a biological arms race. However, the competition among forest trees is primarily, though not exclusively, a contest for sunlight (Harper, 1977; Dawkins, 1987). In each generation, a tree benefits by being, even if only slightly, taller than its neighbors, however tall they may be to begin with. In the course of evolutionary time, successive generations of trees are driven progressively to evolve hundreds of feet beyond their mutual optimum, at staggering costs. Consider, for instance, some obvious items in the list of extra costs:

- *Overload on vertical transmittal*: A sequoia tree has neither a heart nor a brain and, unlike the giraffe, has no need for pumping blood through carotid arteries, but it does have a need for a system of vertical transport and distribution no less challenging. The feat of lifting large quantities of moisture and nutrients through vertical vessels nearly 115 m (390 ft) long, from soil level to the upper trunk and crown foliage, is a bioengineering problem that seems to defy the laws of gravity. The physics of this process is not yet fully understood, but the economic implications in terms of energetic and structural costs are all too clear.

- *Extra structure*: To withstand lateral winds, the sequoia's need for extra trunk volume increases exponentially in proportion to its height; so much so, that the risk of toppling in the wind is nearly equal to the risk of collapse under its own weight.

- *Diminished rate of evolution*: Large size is a recipe for long life, or so it seems, according to one of the best established regularities observed in nature. From bacteria to sequoia, and almost everything in between, life span and body size are highly correlated (see Figure 5.2). There are good reasons for this universal regularity that will not be pursued here except to note that the correlation applies only across species (attempts of prolonging life by gaining weight are thus highly inadvisable to individuals). It should come as no surprise that the life span of the sequoia – the largest living thing (with the possible exception of some underground fungi) – is exceedingly long. According to ring counts of dead or logged specimens, it is clear that the maximum lifetime of the coast redwood exceeds 2,200 years and that of the giant sequoia, an amazing 3,500 years. One can argue endlessly about the merits of a life span verging on immortality. But strictly from an evolutionary point of view all indications point to inexpediency. Long generations put further drag on the rate of evolution (which is already reduced by the fact that large size also entails small populations) and this is, in the final analysis, a price the sequoia as a species has to pay for its height.

The main point here, however, is not only cost but also lack of returns. In terms of access to extra sunshine at the margin of growth the sequoia, in fact, has little to gain (as a population or as a species). Given the (maximum) height of trees that grow adjacent to the sequoia, it can be

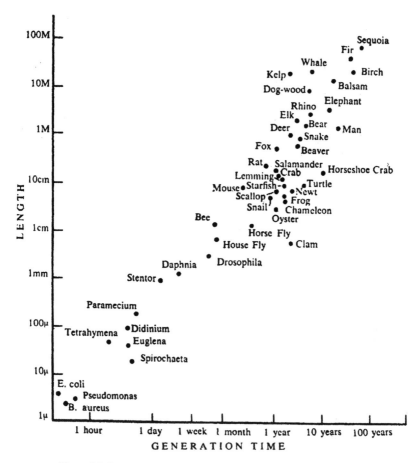

Figure 5.2 **Longevity and length** Generation time plotted against length – a measure of body size – for a variety of animals and plants. The *Sequoia* is at the top of the list on both counts. Logarithmic scales. After Bonner, 1965, Fig. 1.

argued that if all sequoia trees simultaneously were to cut their stature (at great savings) by 1% or 10%, or perhaps even by 30%, their ability to carry on photosynthesis would probably diminish by little or nothing. This serves as an indication to the effect that competition for sunlight, at least in the case of the mature sequoia, is primarily confined to members of the species. The situation is not unusual for a structure produced in a manner of a runaway arms race. We saw the same tendency in the case of the giraffe and, as pointed out by Dawkins and Krebs (1979), this is what we should expect to see under intraspecific symmetric arms races in general. (Granted the ability of niching apart from each other, separate species, or even lineages that ceased to share a common genetic pool, would rather diverge than incur the escalating cost of such races.)

Indeed, the action of an escalating biological arms race for sunlight is probably the most straightforward explanation for the overly extended height of forest trees, notably, the sequoia. However, every complex structure produced by natural selection holds within itself the seeds of alternative (or seemingly alternative) explanations based on their own germs of truth; not the least, proximate causes and a host of incidental selective advantages. If nothing else, certain incidental selective advantages (and disadvantages) always warrant careful examination. How important are these incidental effects in the evolutionary story of the sequoia?

Incidental advantages and disadvantages
Sunlight is not the sole resource for which forest trees compete. Invisible, but equally fierce, is the silent underground struggle for nutrients and soil moisture. Sheer height, on the other hand, is neither the sole weapon nor a sufficient weapon for winning a photosynthetic war. Equally important, if not more so, is the ability to compete successfully in the shade, and withstand fire. For instance, the advance of the rain forest into subtropical woodlands (e.g., in parts of Australia) can be ascribed primary to the ability of the seedlings of tropical species to stand deeper shade than the seedlings of their subtropical counterparts which, on their part, counter by inviting forest fires (see Box 10.1). The sequoia, incidentally, fares quite well on both counts: resistance to fire as well as tolerance to shading.

It should also be noted, that access to sunlight is not the sole benefit afforded by height. Grossly "exaggerated" and exceedingly expensive structures produced by biological arms races, as exemplified by the stature of forest trees (and, arguably, by the human brain) are bound to reveal new unintended applications which can be put to good use. There is no reason not to fully capitalize on a structure, however badly misconceived and expensive, if that structure is already in place and ready to go. The sequoia's outstanding height seems to impart extra benefits (unrelated to photosynthesis) at least in two forms. First, as I have already indicated, there is the element of fire protection. The ability to grow rapidly and establish a canopy aloft a well insulated trunk, as high above the forest floor as possible, is by far the best fire insurance a tree can have. The fact that most forest fires are relatively benign, burning undergrowth and woody debris on the ground but rarely reaching forest crowns, bears testimony to the effectiveness of such fireproofing adaptations. Unlike the rare and highly destructive "crown fires," the far more frequent "ground fires" are actually beneficial to most forest trees. Buried in thick layers of forest litter, sequoia seeds would stand no chance of germination without the aid of such fires.

Another benefit is access to extra moisture which, next perhaps only to sunlight, is the most pressing constraint in the ecology of a tree. There is plenty of atmospheric moisture in the fog ridden range of the sequoia (*Sequoia sempervirens*) – a narrow strip of low land along the northern coast of California reaching just over the border into Oregon.[5] With its crown canopy soaring skyward well into the atmosphere, the sequoia is able to supplement its moisture supply with substantial quantities of water condensed from fog and coast mists.

If the sequoia had a mind of its own it would probably list the acquisition of moisture and escape from fire as equally important, if not more so, than the escape from shade. It would probably also be proud of its stature and be insulted by the suggestion that this lofty structure is in any way "exaggerated" – just as we, humans, are proud of our oversized brain, our ability to do pure math, write poetry, and solve crossword puzzles.

---

[5] Not to be confused with the range of the "giant sequoia" (*Sequoiadendron giganteum*) which is located further inland on the western slopes of the Sierra Nevada. Despite its name, the "giant sequoia" is actually the shorter of the two species: 325 ft (99 m) in height compared with 385 ft (117 m). True to its name, however, the "giant sequoia" is much larger in total trunk volume. Its largest living specimen, the "General Sherman tree," weighs an estimated 6,167 tons.

6　The origins of nepotistic exchange

Without really being aware of it, human beings economically interact under two distinct routines – nepotistic exchange and market exchange – effortlessly switching back and forth between the two regimes. The tendency to turn on and off two separate processes of decision-making (with implications in action and in demeanor) could have been readily diagnosed as some sort of split-personality disorder had it not been so predictable – repeating itself in nearly the same manner in almost any member of the species. The fundamental difference between the market and the nepotistic (or domestic) routines of human activities was examined in the foregoing discussion (especially in Chapter 2) on the grounds of practical considerations and, so far, only with a secondary emphasis on evolutionary ones. The main implication suggests nothing in the way of a behavioral disorder. What it suggests is the coexistence of two distinct adaptations. The main attempts to trace these two adaptations to their separate evolutionary origins are relegated to two chapters in this volume. The present chapter is primarily concerned with the origins of nepotistic exchange. (Chapter 9 will be similarly concerned with the origins of market exchange.)

Primordial exchange at the lowest levels of organization

Certain advantages of division of labor, supported by exchange, occurred at fairly early stages of life on earth. Consider, for instance, the evolution of one of the most comprehensive and amazing systems of exchange and division of labor in nature: a single body of a multicelled organism. Finely tuned exchange among specialized cells, tissues, and organs is inherent in the physiology of such an organism. The "internal environment" is where all these exchanges take place, and that is where all higher animals essentially live their biological lives. The importance of the *milieu intérieur* (French for internal environment) was first fully recognized by a French playwright who turned from drama to medicine at the behest of a literary critic (apparently to the great benefit of both disciplines). This was, of course, Claude Bernard – the undisputed founding father of modern physiology (and a contemporary of Charles Darwin).[1] In Bernard's own words:

> The living organism does not really exist in the *external environment* (the atmosphere, if it breathes air; salt or fresh water, if that is its element), but in the liquid *internal environment* formed by the circulating organic liquid which . . . is the basis of all local nutrition and the common factor of all elementary exchange.[2]

Nobody has ever called into question the advantage afforded by exchange through differentiation and specialization of function and form in the "internal environment," and nobody doubts its antiquity (at least 570 and perhaps 680 million years ago – judging by the age of the earliest known multicellular organism). More ancient and even more fundamental is the molecular exchange among subcellular specialized structures – nuclei and organelles – that carry out specific tasks within the internal environment of each single cell (starting some 850 million years ago with the appearance of the earliest known nucleated cells). Far more ancient yet, according to the symbiogenesis scenario (Margulis, 1981), is exchange and division of labor by "infection" in the form of intracellular symbiosis among prokaryotes (simple cells that lack nuclei) that probably lead to the creation of eukaryotes (complex nucleated cells). This may take us back to the earliest forms of cellular life on earth (perhaps as long as 3.8 billion years ago).

Exchange in the "external environment" between multicellular organisms is essentially the continuation of the same logical process by other means. Social insects and colonial animals represent two distinct extensions in this evolutionary continuum. Mercantile exchange, though unique to human beings, is another extension. With this understanding, there seems to be no shortage of selection pressures that could produce, at almost any stage of evolution, adaptations related to exchange. Natural selection has had ample time (and good reason) to hardwire such

---

[1] In fact, Bernard's best known work, *Introduction à la médecine expérimentale* (*An Introduction to the Study of Experimental Medicine*) was published in 1865, within less than a decade from Darwin's first edition of *The Origin*. The same decade, interestingly enough, saw the publication of the major works by Louis Pasteur and Gregor Mendel. (Mendel's contribution, though, did not receive due attention until 1900, when it was 'rediscovered.')

[2] As adapted from Robin (1979:258).

adaptations into human nature long before the arrival of industrial society. The remaining question is "when" this occurred and "how"?

Convergent body structures

To narrow the search for the origins of exchange, either at the nepotistic level or at the mercantile level, it is useful to make a rigorous distinction between these two levels of organization – this time, purely on the grounds of evolutionary considerations. Treated as an adaptation, market exchange best fits the description of a *specific* character: an attribute of an organism that defines the species or, at least, helps to set the species apart from others. Much of the discussion throughout this book is actually driven by the idea that of the many features that distinguish the human species, market exchange is the most fundamental of all. Specific characters as broad as market exchange that presumably sets us apart from the rest of the natural world, or as narrow as, say, the sixth digit (the "thumb") that sets apart the pandas in the same manner, are generally acquired through *evolutionary divergence*, that is, through a process of selection that accumulates and accentuates differences between species – including speciation, to begin with. This is, apparently, what one has in mind when confronted with the commonplace depiction of evolution as a (genealogical) tree of life: a common ancestor at bottom and new branches that keep budding and splitting off in all directions – growing increasingly removed from their common roots and from each other over time. Divergent evolution is, indeed, the default mode of evolution. It remains in effect unless canceled or overridden under some special set of circumstances by another mode, notably, by *convergent evolution* (to be discussed below).

In sharp contrast, nepotistic exchange more nearly fits the description of a *general* character: an attribute of an organism that makes one species similar in some respects to another, however genetically remote, and however dissimilar they are in all other respects. By its very nature, a general character tends to blur rather than accentuate differences between species. A wing is an obvious example. To the extent that the organ is actually used in flight, it follows common (aerodynamic) principles shared by a wide range of airborne animals (e.g., bats, flying fish, insects and, of course, birds). The design, therefore, cuts not only across species but also across the highest divisions (phyla) of the animal kingdom. In fact, the same design carries over to the plant kingdom as well (since some seeds are duly equipped with wings for wind dispersal). Though nepotistic exchange directly expresses itself in behavior rather than anatomy, as a general character it applies with equal force (and far greater frequency) across the taxa. General characters, especially those associated with expensive structures such as nepotistic exchange (or wings) are typically acquired through a process of *convergent evolution*. Convergent evolution provides a counterbalance that (especially in the case of larger animals) holds in check the bewildering range of diversity that would have otherwise evolved, had the pressures of divergent evolution been left unchecked. As deemed appropriate by the topic at hand, the process of evolutionary convergence will receive a due amount of attention in the discussion throughout this chapter.

Analogy as distinct from homology
In general use, the term *analogy* means a similarity in some respects between things that are otherwise dissimilar. As a form of logical inference, the association of this term with natural history, especially with natural theology, was not always a happy one. Inference from analogy is said to take place when two things known to be alike in some respects are taken – for that reason alone – to be alike in other, as yet unexamined, respects. This form of inference runs, of course, the risk of drawing false conclusions from true premises. A well-known attempt to draw inferences from analogy, with fairly tempting implications for natural theology, is that of the "watchmaker." It was introduced by William Paley (1743–1805) using observed complex designs in nature (starting with the fine design of the observer's eye itself) as evidence to the existence for a Creator. Just as a watch requires a watchmaker, he argued, so nature requires an intelligent designer.

The concept of *analogy* was usefully refined to gain its present meanings in modern biology only when it was set in sharp (and welcome) contrast against the notion of *homology*. The distinction was first clearly made by the British paleontologist Richard Owen in 1843 and was immediately adopted as a standard rule of demarcation in the field of comparative anatomy. Referring to body parts observed in different organisms, the term *homology* was originally reserved for similarities in structure regardless of function (e.g., between the wings of a bat and the flippers of a seal). Conversely, the term *analogy* was reserved for similarities in function regardless of structure (e.g., between the wings of bats and insects). The arrival of Darwinism was greeted with little joy by Owen, but it was the idea of evolution that added to his rule of demarcation its pivotal dynamic dimension. Evolutionists, starting with Darwin, viewed homology as a similarity due to common ancestry; whereas analogy was viewed as a similarity due to adaptation. Under this interpretation, the concepts of analogy and homology could then extend beyond body parts to include physiological processes, and behavior.

What kind of adaptations are bound to produce analogies? The answer to this question is key to the entire process of evolutionary convergence. It will be useful to address the question in a number of steps.

Bilateral convergence

An obvious adaptation that produces analogies is *mimicry*: the superficial resemblance of one organism to another in order to deceive a third. The phenomenon was first detected by the British naturalist and explorer Henry Walter Bates (1825–1892) who spent 11 years (the first two together with Alfred Wallace) collecting specimens in the Amazon basin. Most commonly observed in the insect world, mimicry is primarily a means of deception and protection in predator–prey relationships. For instance, the perfectly edible and harmless viceroy butterfly deceives its predators by mimicking (in coloration) the noxious monarch butterfly. Mimicry is an instance of evolution where the adaptive nature of an analogy is literally evident to the eye, for resemblance, in this instance, is not incidental to the adaptation but its very object. In a paper published only three years after Darwin's *Origin*, Bates squarely attributed the phenomenon (later to be termed *Batesian mimicry*) to natural selection and added in his account a crucial firsthand observation: namely, that the imitating species are comparatively rare in *individuals*, while the imitated abound. The implication is that mimicry, by its very nature, is a form of resemblance that can be spectacular in scope but only limited in overall occurrence.

A "deeper" source of analogies is *adaptive radiation*: the relentless tendency of life to leave no niche on earth unoccupied. Radiation in combination with geographic separation is bound to produce similar structures in unrelated species mainly by the coincidence of similar habitats that are situated far apart (or otherwise cut off). Thus, among Australian mammals, there are marsupial versions of cats, wolves, squirrels, mice, and moles one can hardly distinguish from their "true" placental counterparts in the Old World. Such striking examples of resemblance between species across different orders in the taxa of higher animals provide some of the most graphic proofs of natural selection in action. Similar examples of analogy – or convergent evolution – are found in plants. The New World cacti and the cactus-like African spurges, for instance, belong to separate families in the taxa but are similar in appearance (and function) due to common desert adaptations.

Multilateral convergence
In addition and quite apart from pairwise analogies, as outlined above, there are analogies that occur in a wide range of species simultaneously. These are, typically, produced in response to forces and constraints in the physical environment, that is, in response to abiotic rather than biotic pressures. To see why, it should be noted that the response to the biotic environment (availability of food, predation pressure, symbiotic relationships, etc.) is interactive. It tends to form idiosyncratic narrow niches because it affects peculiar configurations of local species that act on each other with nearly endless permutations. Conversely, the response to the physical environment (climate, topography, availability of moisture, the composition of soils and waters, etc.) is reactive, rather than interactive. It tends to cut across wide ecological habitats, affecting many species uniformly. Hence, attributes of organisms that evolved primarily in response to pressures in the physical environment (e.g., light coloration in snow-covered country) replay themselves in a relatively large number of affected species. As such, they often seem to fly in the face of diversity. Some of these adaptations (including coloration) are fairly shallow in the sense that they are easily reversible or otherwise have little impact on structure and function. Others, however, are fairly deep, affecting in fundamental ways both structure and function. A compelling example to this effect is the streamlined design seen in both fishes and other aquatic forms, as well as in avian forms.

Streamlining is a principle of design that helps fast-moving solid bodies travel with least resistance in fluid. As such, it is the preferred design not only of submarines and aircraft but, of course, also of fish and birds that actually fly. The sleek design is of little or no use to the vast majority of animals that live on the land surface (snakes and lizards, especially of the burrowing kind, are some of the exceptions). Yet, it is essential to some close relatives of such terrestrial animals that at different points in the course of natural history returned to life in water. Aquatic mammals, reptiles, and birds (e.g., the seal, the ichthyosaur, and the penguin – respectively) have converged on this design; so much so, that in some instances an untrained observer can hardly tell the new arrival (e.g., a dolphin) apart from the aboriginal true fish (e.g., a shark). The resemblance that the dolphin bears to the shark goes beyond the acquisition of sleek body contours to include more profound adaptations such as the transfer of the main function of locomotion and propulsion from the limbs to the tail. However, in the final analysis, what matters here most is not the particular resemblance between the dolphin and the shark, but the fact that the same adaptation repeats itself with little variation in all fast-moving animals that operate near the surface in deep waters, regardless of specific ancestry. In this sense, streamlining entails convergence across-the-board.

As is evident from this example, it is occasionally useful to set apart biotic from abiotic causes of evolution and, when appropriate, to single out physical constraints (e.g., the density of water in the case of streamlining) for special attention. If nothing else, physical constraints affect many species simultaneously and, as such, they are more readily bound to impart a wide measure of evolutionary convergence.

Mass convergence

Beyond physical constraints, there is a source of even more universal selection pressures. These are produced by strictly synthetic entities which, by their very abstract nature, are neither biotic nor abiotic, nor do they vary in time and space. At issue are selection pressures produced by invariable mathematical and physical laws of great applicability, by engineering principles of good design, and indeed, by first principles of economic allocation of scarce resources – all produce their own selective pressures. Given the universal applicability of such synthetic selective pressures, we can expect them to produce similar structures in the widest possible range of organisms.

Mathematics, physics, engineering, and economics are not entirely mutually exclusive disciplines when it comes to practical applications. In certain applications they are actually tightly nested within each other. A case in point is provided by the principle of adaptive economies to scale. The evolution of body size seems to follow, under this principle, some fairly predictable rules common to all forms of life, as will be demonstrated below.

Mathematically speaking, all things in nature are inescapably confined to the three dimensional (3-D) Euclidian space, except perhaps the mathematicians themselves. A seemingly simple engineering problem is presented in this space, for instance, by the design of containers for shipping and storage. To economize on packaging material (a waste material) engineers try to minimize the surface-to-volume ratio of a container, among other things, by increasing its size. Doubling its linear dimensions (i.e., increasing length, width, and height by a factor of 2) will quadruple its surface area ($2^2 = 4$), but increase its volume as much as eightfold ($2^3 = 8$). Hence, the surface-to-volume ratio is cut in half, saving 50% in packaging costs per unit of volume. This is a special instance of what economists call *economies of scale*. By the same rule, a threefold increase in linear dimensions will produce nearly 67% savings, and a tenfold increase as much as 90% – and one can go on and on. The bad news, however, is that a tenfold size expansion means no less than 1,000(!) fold increases in weight ($10^3 = 1,000$), which may come as an unpleasant surprise even to the most experienced payload experts. Fully loaded, such an expanded container will gain weight beyond the point of easy handling or sound structural integrity and may soon collapse, one can surmise, under its own bulk. Engineers therefore are faced with a delicate balancing act between economies of scale in surface and diseconomies of scale in weight or, more fundamentally, between an invariable law of solid geometry and the law of gravity.

Designing its own containers in the same 3-D Euclidian space, evolution faces the same balancing act with one important distinction. The outer surface area of any organism, starting with the membrane of a single cell, is not waste material. Quite the reverse, the stuff is probably the most precious structure of the body by the merit of its function and, certainly so, by its sheer scarcity. Apart from its protective role, this structure serves as a monitoring device and a highly selective gateway between the internal and external environments. It controls the exchange of matter (incoming oxygen and nutrients and outgoing carbon dioxide and other waste products) and it regulates the exchange of heat thus helping to stabilize the internal temperature of the living body. Holding constant the rate of metabolism, the load imposed on this structure is proportional to the volume of the body. Its capacity to serve this load is proportional to the surface. Hence, as they grow in size, living bodies are increasingly constricted by an invisible Euclidian straitjacket: the ever-declining capacity of the outer surface to serve additional units of inner volume.

Large organisms are always in dire need of expensive adaptations to compensate for lack of outer surface. Spongy lungs, an obvious (and quite costly) expansion of the outer surface, are found in most land vertebrates. The organ has analogs elsewhere: gills in fish and "book lungs" in some of the larger invertebrates (scorpions, spiders, and snails). Lungs are no longer observed even by remote analogy as one moves to smaller animals such as insects, though in many other ways their bodies are constructed no less elaborately than those of the vertebrates. Smaller in size, they apparently have more plentiful surface area for their volume. The lungs, however, are only one of many structures entailed by the surface-to-volume problem. In fact, the heart and the entire circulatory system down to the kidneys, as well as the gut (especially the small intestine), each in it own right is yet another expensive adaptation to the same synthetic problem in solid geometry.[3] Taken together, the inescapable need to maintain and nourish all these expensive tissues and organs seems to impose considerable anatomical and physiological diseconomies of scale on their large-bodied carriers.

In addition to the internal straitjacket there is an external one: the fact that food is distributed on the surface of land in only two dimensions. The need to nourish a body increases, however, in proportion to its volume in all three dimensions. A fourfold increase in body size (linear

---

[3] It should be noted that, from a physiological and anatomical point of view, the interior space throughout the full length of the gut is an uninterrupted tube that wholly belongs to the "external environment." It is exempt, for instance, from *homeostasis* and all the other strict regularities maintained in the "internal environment." The same is true of the interior space of the lungs.

dimensions) demands, for instance, a daily ranging radius eightfold ($4^{1.5}$ = 8) longer on the part of a terrestrial animal (and a tenfold increase in its size entails a ranging radius more than 30 times longer). The energetic and structural expenditure on locomotion, combined with extra travel time, exponentially increase with body size and rapidly become an additional limiting factor to reckon with. The Euclidian straitjacket, so to speak, is double-breasted.

One can go on and list a myriad of other adaptations – notably, adaptations to gravity – all of which imply further diseconomies of scale. Best testimony to this effect is provided by the fact that an ant can lift and carry with ease a load many times its own weight, but a mule, our most efficient pack animal, is unfit to carry on its back even one of its own. One way or another, size operates to the disadvantage of the sizable. Indeed, viewed purely from a physiological standpoint almost all organisms with few exceptions seem to grow in size to a point where all economies of scale have already been exhausted and diseconomies of scale set in. The only exception where body size imparts a clear physiological advantage, as far as I am aware, occurs in certain adaptations to cold in extreme frigid environments and, even there, only under special conditions (i.e., only to the extent that the risk of hypothermia steadily dominates the risk of hyperthermia in all seasons). At the same time, to be sure, no serious students of the "economy of nature," from Aristotle to Linnaeus to Darwin and beyond, ever in the least doubted that all living things, each in its own peculiar way, are nearly always just the right size for their own good, however one wishes to define it – phenotypic survival or genotypic fitness. Is there a contradiction?

A (seeming) contradiction quite similar, in a way, frequently occurs to economists concerned with the issue of optimal firm size. To operate efficiently in an economic sense, a business firm must keep growing in size not to the point of minimum *per-unit* cost, as one might initially surmise, but actually beyond that point, sometimes even well beyond it, and almost never short of it. The most simple, albeit partial, explanation for this strange phenomenon can be traced to the fact that no firm in the business world can indefinitely, and costlessly, increase the number of plants and factories under its operational control. Sooner or later it is bound to exhaust, among other fixed factors, its managerial capacity –

for managers, like all humans, have only limited capabilities. For that reason, as they grow in size, firms tend first to expand their existing plant (or plants) well beyond the margin deemed otherwise optimal and only then consider launching new ones. The constraint on expansion in the extensive margin (i.e., in the number of plants) leads to overexpansion in the intensive margin (in plant size). Given a fixed limiting factor (such as managerial capacity), this is precisely what one might expect under the economic *law of diminishing returns* in production. In general, however, this "law" is applicable to all entities composed of fixed and variable constituents in variable proportions. Among other things, it regulates the optimal size calibration not only of firms but also of living things.

The law of diminishing returns applies to the natural world with added force, for this is the world where expansions are far more confined by fixed limiting factors than in the business world. A firm may still operate in a number of plants but no organism can live its natural life in more than one distinct reproducing body (excepting incarnation). Making the best of a scarce resource, living things are driven by pressures in the environment to push, literally speaking, the envelope. Given a fundamental tradeoff between environmental opportunities and physiological needs, evolution is bound to strike a balance at the point where *in vivo* bodies systematically exceed their, otherwise, optimal *in vitro* size. The implication is that living things rarely minimize energetic and structural cost per unit of biomass. This is apparently true not only of the larger macroorganisms, like an elephant or a mule, but also of the smallest microorganisms down to the cellular level.[4] However, the brunt of the cost is disproportionately borne by the largest.

Large size imparts uniform pressures that call for equally uniform adaptations and, eventually, for an increasing measure of convergence among those most affected by it. As we saw, size entails expensive adap-

---

[4] Strictly to minimize (intracellular) transport costs cells can be expected to take on a spherical shape unless otherwise instructed by an overriding function. In reality, however, cells are rarely observed to take this shape except when they are about to divide. Free living (or disembodied) cells tend, in fact, to assume shapes that more nearly maximize their surface-to-volume ratio even at considerable extra cost on intracellular transport. The fact that our tiny red blood cells are normally doughnut-shaped suggests to me that even the smallest of the human cells is already oversized.

tations (such as lungs) on the part of large animals, a burden from which their smaller counterparts are exempt. Facing a common liability, they are expected to converge on common solutions at the cost of custom-made idiosyncratic ones. The latter are left to small animals – the specialists in peculiarities. The expected pattern implied by these considerations is clear: convergence in large animals and diversification in small ones. This theoretical expectation is not entirely an idle speculation. Indeed, a salient observation in nature is that very small plants and animals are far more diverse than very large ones. Among the 4,000 species of mammals found throughout the world, points out E. O. Wilson, "a thousandfold decrease in weight means (very roughly) a tenfold increase in the number of species. This translates to about ten times as many species the size of mice as species the size of deer" (1992:207). Beyond the number of species in each size category, there is also the range of body size over which genera, families, and higher taxa are distributed. For instance, among non-mammals on land there are many species of reptiles and birds and some invertebrates that are the size of mice, but only a few the size of deer, and none is greater or even near the size of the zebra or mule. With the dinosaurs long extinct, the gift (or curse) of size seems now to be the exclusive dominion of the mammals. To me this sounds much like explosive diversity in the small and blanket convergence in the large.

All in all, it is only fair to conclude from the discussion so far that synthetic constraints, however invisible, are a consequential force in the evolution of morphological structures – especially, in the convergent evolution of such structures. The extent to which similar convergent pressures transcend also to behavior and social structures is a question to be addressed in the following discussion.

Convergent social structures

Engineers, Richard Dawkins points out, "are often the people best qualified to analyse how animal and plant bodies work, because efficient mechanisms have to obey the same principles whether they are designed or designoid" (1996:19). The question is whether the same logic can be extended one order of organization higher – from body structures to social structures. There is no shortage of synthetic principles that can explain social behavior. Chief among them, it seems, is the principal of division of labor. There is thus no reason why we should not try to approach social adaptations with the same degree of scrutiny we applied to anatomical and physiological adaptations in our earlier discussion of body size and streamlined design. At issue are adaptations that repeat themselves in widely separate species and, for that reason, can be clearly ascribed to analogy (as distinct from homology); indeed, to convergent evolution in social structures.

One of the most compelling arguments for the existence of evolutionary convergence in social structures was made by E. O. Wilson on the basis of the following observation:

> Among the deepest and therefore most interesting cases of convergence in social behavior is the development of sterile worker castes in the social wasps, most of which belong to the family Vespidae, and in the social bees, which have evolved through nonsocial ancestors ultimately from the wasp family Sphecidae. The convergence of worker castes of ants and termites is even more profound . . . their phylogenetic bases are considerably farther apart: the ants originated from tiphiid wasps, and the termites from primitive social cockroaches. (1975:25)

Based in part on new discoveries, one can extend the analogy, and thus the incidence of convergent evolution, all the way to the mammals and beyond. For instance, a social structure resembling that of insect colonies was observed in the naked mole-rat (*Heterocephalus glaber*). On inspection (see Box 6.1), four key structures of division of labor known from the social organization of the social insects are displayed by this burrowing rat-like mammal: (1) breeding is primarily limited to a single female, or queen; (2) care of the young is cooperative; (3) division of labor exists between workers and soldiers; and (4) further division of labor takes place among workers assuming specialized tasks in subsistence. Taken together, these four structures amount to *eusociality*. The pattern repeats itself, to a lesser degree and in muted forms, in other social species among the mammals, especially in certain species of the *Canidae* family (e.g., the African hunting dog as noted in Chapter 2). Moreover, there are recent indications for the possible existence of eusocial species not only among insects and mammals, but also among certain forms of marine invertebrates.[5] The fact that *eusociality* repeats itself at different crossroads in the taxa despite considerable intervening phylogenetic distances is a fairly strong additional indication for convergence in social evolution.

Four distinct pinnacles of social evolution have been emphasized by Wilson (1975) as they are typically observed in (1) colonial invertebrates; (2) the higher social insects; (3) vertebrate societies; and (4) human beings. The specific patterns and the underlying adaptive mechanisms are primarily determined, no doubt, by phylogenetic descent. However, the general pattern bears testimony to the action of convergent evolution. It follows that these pinnacles are no discrete occurrences but more nearly a continuum of wide-ranging responses to a common selective advantage, most likely, the advantage of division of labor.

We humans actually occupy two distinct points in this continuum. First, in our capacity as agents who operate in market networks we probably defy any known precedent in our immediate primates order, if not in nature at large. However, in our capacity as agents who interact in nepotistic networks we seem, zoologically speaking, to fall well in line with the mammalian heritage. Despite symbolic language and vastly improved communication skills, relatively little in the way of innovation is displayed by us in this respect. The dichotomy is perhaps subliminal in perception, but in actual behavior we hardly ever fail to compartmentalize our action in line with it. The largely successful dual attempt to avoid nepotism in business and commercialization in family affairs, as I already indicated, is in itself an important defining quality of human nature. When we explicitly recognize this dichotomy and consider the two branches of human behavior each apart from the noise of the other, we should stand a better chance to obtain a far more coherent picture of our relation to our closest evolutionary relatives – the great apes and other primates.

---

[5] The discovery of the first known case of eusociality in a marine invertebrate, I was informed, was made as recently as 1997 by J. Emmett Duffy of the Virginia Institute of Marine Science. The species, a sponge-dwelling coral-reef shrimp (*Synalpheus regalis*) live in colonies of more than 300 individuals with a single reproductive female functioning as a queen, while multiple generations of her nonbreeding offspring serve to protect the colony against intruders.

Box 6.1 The naked mole-rat: a mammal converted to *eusociality*

A small hairless rodent native of East Africa, the naked mole-rat (*Heterocephalus glaber*) lives in underground colonies (up to 100 strong), each with a single queen (the largest colony member), two or three breeding males, soldiers, and workers. The youngest workers, and hence the smallest, are primarily engaged in the collection of food (mainly from tubers and other plant roots). As they grow bigger, they spend more time digging foraging burrows and other tunnels, which is their most energy consuming and finely coordinated team activity (see Figure 2.1). The largest non-breeding members serve as soldiers who assist in the care of the young. Size and age thus determine the division of labor, as is true generally of the social insects.

Most or all reproduction in a colony of naked mole-rats is by the queen who also suckles her litters, but the workers tend and feed the younglings once they are weaned. All mole-rats in a colony are able to breed, but the queen controls their behavior with pheromones, which repress their normal reproductive development and instincts. In addition, much like bumblebee queens, she controls her daughters by aggression, attacking them whenever they attempt to reproduce on their own. Consequently, like social insects, mole-rats are characterized by self-sacrificing sterile castes, while a low-keyed struggle continually takes place between queen and workers for the opportunity to reproduce. The same kind of low-keyed conflict is known to exist in colonies of certain ants, bees, and wasps. When the queen dies, several contending females grow rapidly. One eventually intimidates the others and becomes the new queen. Again, a similar contention for the alpha position takes place in certain wasp and bumblebee societies when their queen is removed and some of the workers fight amongst themselves for the right to replace her.

The primate connection

The central organizing principle in the social life of all mammals is the relationship between the lactating female and her offspring. All other arrangements are derivatives of this invariant and most binding constraint on resources, time, and effort. The fact that female mammals nurture their embryos internally and nurse their infants externally clearly defines the female role in parental care and, by extension, predetermines the sexual division of labor in general. Secondary aspects of parental care, and the entire mating system, are more anecdotal to the particular species depending primarily on environmental constraints such as the availability and the distribution of food sources, predation, and the physical elements. Indeed, zoologists have long recognized the fact that female mammals tend to distribute themselves primarily in relation to the distribution of food, whereas males distribute themselves primarily in relation to females. This is the general background on which nepotistic networks (i.e., mating, parental care, and the corresponding kin and social structures) are established among the mammals, and primates are no exception.

Given the invariant factor, the mother–offspring bond, all the other components of parental care and the entire mating system are fairly variable not only across separate species but, more interestingly, also within species. The issue has been examined in a comprehensive study by Dale Lott (1991). Based on several hundred vertebrate species, the study classified each by its characteristic mating system and parental-care system. Monogamy, polygyny, promiscuity, polygynandry, serial polyandry, and simultaneous polyandry were the alternative mating systems he used in his classification. Two-parent, one-parent, brood split, communal care, and helpers at the nest were the care-giving alternatives. An additional factor of special interest to primatologists is the phenomenon of adoption of orphaned young by related and unrelated adults, a behavior pattern in certain species of primates.

Intraspecific variation, the central point in Lott's work, raises perhaps the most interesting question about such nepotistic networks in animals: the question of social plasticity. How soon and how far do nepotistic systems change as a species adjusts to new conditions? Pertinent conditions fluctuate in time and space, and across different populations of the same species. Can we expect to see perceptible adjustments in one generation or do we have to let evolutionary time take its course? Lott's overall findings suggest far more plasticity than an occasional observer might suspect. Consider some implications for the primates in the *hominoidea* superfamily that includes apes and humans.

There are four existing main groups of apes in this superfamily, excepting humans, each with a distinct nepotistic network of its own. The gibbon was first to branch off from the hominoid lineage (some 20 million years ago by DNA hybridization dating).[6] The common gibbon (*Hylobates lar*), one of about half a dozen species of gibbons, is quite a typical representative of the group in that it is largely monogamous. These small white-handed Asian apes live in solitary two-parent reproductive units of two adults and several young. The male and female share nearly equally in the effort of parental care. Male gibbons have been observed to take over the female role in tending their orphaned young. Next to diverge (about 15 million years ago) was the solitary orangutan. The orangutan live in small single-parent groups of one or two mothers and offspring. Adult males join them only on a temporary basis for mating, but live alone the rest of the time – an adaptation that helps reduce feeding competition with their own offspring and mates by using separate feeding sites and lower canopy levels than females and their young (Clutton-Brock, 1977). The relatively large (75 kg) orangutan male, apparently, consumes too much food to afford the same kind of close monogamous pair bonding enjoyed by the much smaller (8 kg) common gibbon. Partly as a consequence of this arrangement, the mating system is determined by overlap in feeding territories resulting in loose polygyny to promiscuity.

The third group, the gorilla (diverged some 10 million years ago) represents a strict polygyny system of mating. Unlike gibbons and orangutans, they spend most of the time on the ground. Gorillas live in single family troops (of up to 30, but more typically 10) that may include in addition to the mothers and their offspring one or two young adult males, all under control of a dominating male, the silverback. Parental care is thus characterized by the presence of both parents with considerable time and indulgence afforded to the young not only by their mother but also by their father. There are also reports of silverbacks taking care of their orphaned offspring.

Finally, the chimpanzees and our prehuman lineal ancestors reached a fork in the road of evolution and went their separate ways (only about 5 million year ago). Today, the chimpanzees live in loosely knit highly interactive and communicative troops two or three times more numer-

---

[6] An account of the DNA hybridization technique and the dates for the divergence of the hominoid lineage obtained from it, as well as from fossils, is provided in Sibley (1992).

ous than the gorilla troops. The mating system is essentially promiscu-
ous and, consequently, the single mother mode of family life is the basis
of parental care. Unlike the solitary style of the orangutans, single par-
enthood in chimpanzees is performed in a commune context with
intense interaction between the mothers and other adults, among the
young, and between the young and their older siblings. The strong matri-
focal bond continues throughout the life of the mother, as do sibling ties
throughout sibling lives. Orphaned juveniles are typically adopted by
their older sisters.

Human beings display a far larger degree of intraspecific variation and
two major, partly interdependent, adaptive innovations: (1) direct feeding investment in mates and offspring by the male, as well as (2) loss of
overt estrus and concealment of ovulation by the female. These two innovations are adaptively traceable, perhaps, to the human departure from
the primate feed-as-you-go strategy of subsistence to hunting-gathering.
It should be noted, however, that neither this nor all the ensuing transi-
tions in the course of evolution led humans to abandon any cardinal part
of the primate repertoire of reproductive behavior. On the contrary, occasionally and interchangeably humans use almost all the reproductive
strategies separately observed in the apes, except perhaps for the systematic promiscuity of the chimpanzee. Some mating patterns practiced
by peoples in all parts of the world since ancient times do not exist in the
apes but occasionally are found in other primates (e.g., cooperative
polyandry, which has not been found to occur regularly in any other
species of mammals except for humans and certain species of South
American tamarin monkeys (Goldizen, 1990)). Consequently, beyond the
two innovations mentioned above, one way humans differ from extant
apes in terms of reproductive behavior seems to be the matter of their
sheer flexibility across patterns separately observed in different ape
species but, simultaneously, occur in none.

It can be argued, of course, that the lack of (intraspecific) flexibility
seen in the mating systems of the apes is merely an artifact of human
perception. The fact that the range of data on humans under our direct
awareness far exceeds the data on nonhuman societies invites a difficulty of asymmetric information. Given the sheer quantity of information, one is more likely to encounter peculiarities and get the impression
that a well-observed human phenomenon is more diverse than its poorly

observed counterpart in an animal. In time, this difficulty tends to be partly rectified as new reconnaissances are undertaken and new observations are made. For a long time, for instance, the common gibbon was viewed as the very model of a monogamous mammal in terms of mate fidelity. Minimal dimorphism, the fact that pairs spend most of the day in close proximity in a relationship that apparently lasts over many reproductive cycles, if not for a lifetime, and the fact that the male is actively involved in caring for the young (and in "guarding" their mother from approaching neighboring males) – all helped to boost up this stereotype. More recently, however, the stereotype was partly shattered by new observations on the actual conduct of these Asian apes in the wild. Following three study groups of white-handed gibbons (*Hylobates lar*) in the rain forest of Thailand over two separate periods (totaling about 20 months) researchers were able to detect extra-pair sexual activity with some regularity. Despite the fact that gibbon females were rarely found without the company of their pair-males, such acts of sexual infidelity occurred at an estimated frequency of 12% (Reichard, 1995). Interestingly enough, of the 7 extra-pair copulations (out of a total of 66) reported by Reichard, at least in one instance the act was apparently performed while the female was already pregnant. This may indicate that the practice is adaptive not only from the viewpoint of the female but, to a certain paradoxical degree, also from the viewpoint of her regular consort. It reduces the probability of infanticide by neighboring males. If the original male were to die, or otherwise to be replaced, its offspring would stand a better chance of survival. Further observations (especially new data from DNA tests of kinship) will probably prove that the repertoire of reproductive behavior in the gibbon and other apes is more diverse and more subtle than we now commonly believe. Yet, it is still safe to assume that even when all the data are in, no single species of apes will ever display a system of mating nearly as diverse as humans experience across cultures, or even within a culture.

If indeed the reproductive systems and, consequently, other social structures in the apes are less plastic than in humans, then one may reach the conclusion that the four groups of extant apes may represent radiation from the main *Hominoidea* lineage into narrow niches in the environment that are liable to produce highly specialized, if not endemic, species. This conclusion is partly supported by findings from the field of primate paleogeography as noted, for instance, by Lovejoy (1981:347):

> The great apes are markedly restricted and occupy only minor areas where minimal environmental changes have taken place since the early Miocene. Yet the fossil record shows that their lineal ancestors (dryopithecines, *sensu lato*) spread throughout the Old World following the establishment of a land bridge and forest corridor between Africa and Eurasia about 16 to 17 million years ago, and that they enjoyed considerable success after their colonization of Europe and Asia.

If this is the case, then humans – at least in their capacity as nepotistic animals – may in the final analysis be a more authentic representation of the common ancestral type than any of the other existing species among the hominoid primates. In any event, the origin of nepotistic human behavior, as distinct from market behavior, can be clearly traced to the larger pool of primate inheritance.

Humans could hardly find, however, a point less promising than the primate pedigree from which to start their descent into the market economy. Primates are unimpressive actors in the exploits of division of labor, even by mammalian standards, and poor manipulators of their environment as such (e.g., compared with burrowing animals or the beaver). Except for inadvertent dispersal of seeds, they show little interest in developing symbiotic relationships within or without their order. Primates also show no inclination for geographic mobility nor for hoarding food over time. To the extent that they use food storage, they do so mainly in order to hide it from other individuals who may expropriate or steal it (Vander Wall, 1990). In these, as in other factors essential to the future development of the human economy, the primates had little to offer.

Given this humble beginning in the primates order at large, being related more closely to the apes than to other anthropoids was an advantage nonetheless . The distinguishing morphological feature of the ape – versatile use of freely swinging forearms – is one important gift humans could not acquire from the quadruped monkey. There are also all the benefits that go with the larger brain of the ape: rudimentary use of tools, improved communication skills, life-long recognition of relatives and neighbors, and so on. What human beings could acquire neither from the monkey nor from the ape were the skills of market exchange which, in the present analysis, are the most important feature of all. In the acquisition of these skills, humans were largely left to their own devices.

7 Baboon speciation versus human specialization

The chimpanzee is our closest nonhuman relative and, on the whole, is probably the best animal model of human *individual* behavior. A society of chimpanzees, however, is not necessarily the best model of human society. Strictly in terms of *group behavior*, humans may bear – perhaps uncomfortably – a closer resemblance to a society of baboons. In fact, of all the animals that live in groups, only the baboons provide a model of group formation (known as the *fusion–fission system*, after Kummer, 1997) that defines the very concept of a *group* in a sense applicable to human society. Ironically, the baboon is a quadruped dog-snouted primate that on a phylogenetic scale occupies a place nowhere as nearly related to humans or chimpanzees as both are related to each other. Darwin was keenly aware of this fact but, nonetheless, made nearly twice as many separate references to baboons than to chimpanzees in the *Descent of Man* (see Box 7.1). Almost a century later the importance of the baboon's ecology and society to the study of human evolution (especially in its Australopithecine stage) was recognized in the literature following an influential paper by DeVore and Washburn (1963). The present discussion further follows the evolution of the baboons in an attempt to better understand some conceivable early indications for the existence of human exchange. On this issue and a few others (e.g., sex-roles and warfare) I have found the insights gained from the natural history and social structures of the baboon to be more helpful than comparable insights gained from the chimpanzee or other apes. Perhaps we have something in common with a monkey that we do not share with an ape.

Parallels in the feeding ecology

The unique evolutionary experience humans share with baboons, and with no other family of large primates – apes or monkeys – is a parallel ecological conversion. In a fundamental transition for both, grass replaced the forest tree as the central feature of subsistence. The ancient association of humans with relatively open country is evident in faunal analysis and fossil pollen analyses showing that the earliest known fossil specimens of hominids between 4 and 2 million years ago all derive from strata that were laid down under nonforest conditions (Isaac, 1983). To this day, our menu is 50% or more derived from grasses (e.g., wheat, rice, corn, barley, sugarcane, etc.), or from the products of grazing animals and animals that convert grain into protein (e.g., pigs and chickens). The share of grass in the baboon's food intake is on average not much differ-ent. In terms of diversity, composition, nutritional balance, and – above all – the capacity to secure all the basic ingredients in the wild, the baboon diet bears a strikingly close resemblance to the human diet. Primarily vegetarians, baboons consume plant food with an emphasis on grass in virtually all edible forms: roots, stems, leaves, flowers, and seeds. Much like humans, however, they regularly supplement plant food with high protein animal food, including a wide variety of sea food acquired, for instance, in tidal pools along shorelines. In fact, they cross rivers and submerse themselves in water (especially on hot days); so much so, that in a way they show a degree of fascination with aquatic activities well known among humans (Morgan, 1994), but not among chimpanzees. Interestingly enough, baboons relish alcohol with humanlike enthusi-asm, and seem to suffer from hangovers as well (Box 7.1).

Box 7.1 Darwin and the baboons

Curiously, notwithstanding the fact that by descent humans are related to apes far more closely than to other primates, in the *Descent of Man* Darwin refers more often to baboons than to chim-panzees. My rough count shows at least 17 separate instances of reference to the former compared with 9 to the latter. Nobody can suspect Darwin of being incognizant of mankind's place in the primate order. The extent of his awareness regarding the phylogenetic proximity of humans to different families of pri-mates is evident in his willingness to accept T.H. Huxley's then (if not now) sweeping proposition "that man in all parts of his organization differs less from the higher apes, than these do from the lower members of the same group" (Darwin, 1874:153). Huxley reached this conclusion on the grounds of comparative anatomy. Essentially the same conclusion was reached in the first breakthrough of modern molecular systematics when

Morris Goodman (1962) clearly demonstrated that humans, chimpanzees, and gorillas are genetically hardly distinguishable and, as such, are more closely related to each other than any is to an orangutan - not to mention the baboon.

Darwin's "biased" attention can be attributed, in part, to the availability of data. Unlike the restricted distribution of the chimpanzees, the widespread distribution of the baboons throughout the continent of Africa often put them in close quarters with local human populations and, thus, under constant collateral human surveillance. Written accounts date back to ancient Egyptian hieroglyphs which often depict the "sacred baboon" as an object of worship and in more mundane roles as well (e.g., helping its human handlers to pick figs). Darwin relied more heavily, however, on written reports by the nineteenth-century German zoologist and ornithologist Alfred E. Brehm (1829–1884) who served at the time as director of the Zoological Gardens of Hamburg and published several books summarizing his extensive field observations in Ethiopia and other parts of northeastern Africa. Though highly anecdotal and occasionally quite anthropomorphic, many of Brehm's accounts (especially those cited by Darwin) were authenticated by future research and, in time, proved to be fairly insightful.

In no little part, I think, the baboons got Darwin's (and Brehm's) attention also because they offer irresistible analogies with humans. Listed below are but two of Brehm's accounts as relayed by Darwin. Both bear upon the discussion in this chapter or, at least, are tangential to it. The first account is concerned with the effects of alcohol:

> Brehm asserts that the natives of north-eastern Africa catch the wild baboons by exposing vessels with strong beer, by which they are made drunk. He has seen some of these animals, which he kept in confinement, in this state; and he gives a laughable account of their behaviour and strange grimaces. On the following morning they were very cross and dismal; they held their aching heads with both hands, and wore a most pitiable expression: when beer or wine was offered them, they turned away with disgust, but relished the juice of lemons. (1874:7)

Citing another report by Brehm, the second account bears more directly on the antipredatory behavior typical of the baboons:

> In Abyssinia, Brehm encountered a great troop of baboons who were crossing a valley: some had already ascended the opposite mountain, and some were still in the valley: the latter were attacked by the dogs, but the old males immediately hurried down from the rocks, and with mouths widely opened, roared so fearfully, that the dogs quickly drew back. They were again encouraged to the attack; but by this time all the baboons had reascended the heights, excepting a young one, about six months old, who, loudly calling for aid, climbed on a block of rock, and was surrounded. Now one of the largest males, a true hero, came down again from the mountain, slowly went to the young one, coaxed him, and triumphantly led him away – the dogs being too much astonished to make an attack. (1874:102)

Indeed, group defense is used quite successfully by troops of baboons to fend off common predators far more formidable than dogs – leopards and occasionally even lions. Roughly equal in body size and endowed with canines perhaps twice as large as a dog, the "hero" of the story was apparently unimpressed by its canine attackers.

Perhaps not by coincidence, a baboon was chosen as donor in the first attempt to transplant an animal liver into a human. The surgeons at the University of Pittsburgh Medical Center (where the operation was performed on June 28, 1992) apparently judged the baboon liver to be the most compatible replacement. From an evolutionary point of view this choice should come as no surprise. The limiting factor in any feeding adaptation is not only the ability to acquire and digest a certain configuration of staple ingredients but also, if not more so, the ability to remove typical toxins and harmful substances – the primary responsibility of the liver. As part of the digestive system, the liver processes products of digestion carried from the gut (through the portal vein). Its main function is to metabolize, store, and detoxify nutrients, as well as to remove waste materials. Plants have evolved toxins to fend off animals that feed on them. Pastoral images notwithstanding, the grazing and browsing ecology is in fact one of the fiercest battlefields in nature. Batteries of toxins are the major weapons deployed by plants. A major counter-weapon at the disposal of animals is the liver. The design of the liver must therefore reflect to some degree the typical configuration of toxic chemicals ingested, as is true of more elementary digestive organs with respect to other peculiarities of the diet.

Just by looking at the gut of an animal, experts can make an educated guess about its routine diet. As already indicated (in Chapter 5), the human gut is quite unique. When the relative proportions of the human gut are compared with those of other primates, it is difficult to find a good match (Milton, 1987). The human gut is dominated by the small intestine, whereas most anthropoids (including the apes) show notable volume in the cecum or colon, or have a highly specialized stomach. For instance, the small intestine accounts for roughly two-thirds of the gut volume in humans, but for less than a quarter in chimpanzees and less than one-sixth in gorillas.[1]

One of the rare exceptions Milton (1987) mentions is, not surprisingly, a savanna baboon (the subspecies *Papio papio*) which seems to come close to humans with its relatively enlarged small intestine compared to other sections of the gut.[2] I am not aware of comparable measurements taken for other subspecies (or closely related species) of the savanna baboons, but it is safe to assume that any deviation across subspecies (or across interbreeding species) could not be substantial. The similarity in relative gut proportions of humans and savanna baboons is not derived, of course, from a close common ancestor. Rather, it appears more nearly to represent an instance of evolutionary convergence due to an overlapping past, if not present, common feeding ecology. Indeed, as Chivers (1992) notes, of the factors that determine the patterns of primate guts, the direct functional adaptation of each species emerges as more important than phylogenetic affinity.

One aspect of diet that does seemingly set the baboon apart from morphologically modern humans is insectivorousness. Insects and other

---

[1] Based on data presented by Milton (1987).
[2] The only other exception mentioned by Milton, in this context, is the New World capuchin monkey (*Cebus* spp.).

invertebrate prey serve as a major source of high quality protein for the baboon – as they probably served our Australopithecine ancestors (Zihlman, 1983). This difference can be explained, in both cases, by relative body size. The baboons are about the size of the early Australopithecines (i.e., roughly half the size of modern humans).[3] Between vertebrate predators and prey, a primate of this size is more likely to end up as dinner than as diner. Grasshoppers, spiders, scorpions, worms, and other invertebrates are fair game that the baboon can scarcely afford to ignore – nor could such a selection be easily ignored by an ancient hominid as small as "Lucy" (*Australopithecus afarensis*) or, perhaps, even *Homo habilis*. Incidently, shell fish and other aquatic invertebrates are to this day a welcome dietary supplement to modern humans and, in certain regions, land insects still serve as staple food.

On their part, baboons occasionally hunt small vertebrate prey: birds, birds' eggs, and mammals up to the size of hares or fawns. In this respect they are highly opportunistic. For instance, in response to an increase in the antelope population in Kenya (due to human annihilation of large predators) baboons markedly increased the frequency and sophistication of their own predatory activity (Harding and Strum, 1976). In an encounter with loosely protected livestock, they may rise to the occasion and devour domesticated animals the size of young sheep or goats. They do not spare, of course, farm produce in other forms – a fact that helped them gain a reputation as agricultural pests. Indeed, partly due to agricultural expansion by humans, the baboon is presently the most successful primate as measured by geographic distribution – not counting humans themselves. Moreover, while most primates are threatened by deforestation, certain species of baboons – certainly the savanna baboons – actually benefit from it. The common adaptation to life in grassland apparently linked the baboon to humans in a fairly prosperous but, environmentally, not entirely holy alliance.

Antipredator behavior

The human parallels with the baboon go much farther than diet. The

---

[3] Depending on species and gender, they measure about 30–90lb (14–41 kg) in weight and 20–45 inches (50–115 cm) in length, not counting the tail. Like the Australopithecine, male baboons are roughly twice the size of females.

transition from woodland to grassland is more than merely an adaptation of an arboreal animal to terrestrial browsing on forest floors or forest clearings. Life on grassland entails daily foraging expeditions far from the safety of trees. Daily ranging of 5–10 km, depending on the density of the vegetation, is essential for baboons living in typical savanna. Shorter trips may suffice in richer semi-forest habitats. But trips almost twice as long (9–19 km per day) are required to sustain the hamadryas baboons in their semi-desert treeless range, where the vegetation is sparse and refuges (typically, rock walls inaccessible to predators) are few and far between. The main challenge is exposure to large carnivores.

A single baboon, especially an infant or a female (half the size of a male), that ventures away from the safety of trees or cliffs stands little chance of escaping an attack by a leopard (their main predator) or hyenas. The most obvious adaptation for contending with this kind of danger is group defense. On open grounds baboons face their enemies in battle formation. Males, and occasionally females, join together in a coordinated defense of the troop – mobbing smaller predators and fending off larger ones – or, as a last resort, forming a rear guard to troop retreat (see Box 7.1).

In itself, coordinated group defense is most effective against carnivores that hunt by stalking prey (e.g., wild dogs and hyenas) and less effective against those who hunt by stealth and ambush (e.g., big cats). A leopard or lion in hiding can make a quick snatch of a youngster and retreat with impunity long before collective lines of defense are drawn by the troop. To cope with the threat of such stealthy attacks, baboons have developed additional adaptations designed to lower the risk of being caught off guard, especially on open grounds. DeVore and Washburn (1963:343) point out that the behavior of baboons in a wooded area contrasts strikingly with their behavior in the open: "A baboon troop that is in or under trees seems to have no particular organization, but when the troop moves out onto the open plains a clear order of progression appears." Crossing open ground, baboons will usually travel in highly regimented large formations. Females and young surrounded by the strongest high-ranking males are lined up at the center of the procession. Lower-ranking young males provide front and rear guards to the marching party (Figure 7.1). In repeated trips between two points in their range, baboons constantly change the configuration of trails on which

Figure 7.1 **Baboons in battle formation** Any resemblance to infantry on patrol is (perhaps) coincidental. The images in this picture actually belong to an offshoot of a band of hamadryas baboons before it embarks on a daily foraging trip. Adapted from Kummer, 1997.

they choose to travel, thus, making their coming and going less predictable to an ambushing predator (Kummer, 1997:63−64). The overall ensemble of such defensive mechanisms apparently works fairly well.

The extra risk baboons take by venturing into open grassland is apparently not matched by a corresponding increase in the rate of predation, as might otherwise be expected. In fact, according to Estes: "nearly all baboonologists have commented on how little predation they have observed" (1992:518). This achievement was accomplished primarily through refinements in behavior and communication, and a highly complex and variable social organization reaching a pinnacle with those baboons who must cope with desert life (i.e., the *hamadryas*).

Of immediate interest, however, is the extent to which predation pressure in open country compels a free spirited primate, when away from shelter, to operate in highly disciplined collective formations at what seems to be considerable loss of individual freedom to take independent action. A pregnant female hamadryas baboon, for instance, can ill afford to enter labor in daylight when the troop is on the move or foraging. She will usually give birth only at night when the troop is sheltered in its sleeping quarters (Kummer, 1997: 204−5). On the other hand, as pointed out above, regimented troop formations typical of baboons on open ground tend to break down as soon as they move under trees or otherwise near shelter from predators. This degree of behavioral flexibility in response to changing predation pressures has analogies in human evolution and some bearing on the development of market exchange.

For much of their tenure on earth, early human and prehuman societies faced essentially the same carnivores, and more formidable ones that are now extinct, on similar if not the very same open-country grounds occupied today by the baboons − with little or nothing more in the way of anti-predatory defenses or shelters. Some 3 million years of (*Australopithecine*) savanna survival preceded the first appearance of the hand ax, a replacement for canines, and 2 million additional years for the arrow and bow. In the meantime, the only conceivable safeguard against predators was collective defense in close group formations. If there are any doubts, in this respect, just remember our fascination with marching bands and with parades and ceremonial processions of all sorts and in almost all walks of life, from religion to politics to education − coronations, commencements, weddings, and funerals − and, not the least, in military affairs. Human beings no longer wage all-out wars against predators, but occasionally we fall back on the same ancient techniques against each other. From the ancient phalanx to this day, maneuvers in close battle formations on open grounds always dominated the theaters of war and, in peacetime, they dominate all sports events in a stadium. Yet, by no coincidence, one walk of life where organized public congregations and processions have no particular role to play is the marketplace. The business world is largely organized on the principle of unilateral decisions and bilateral transactions – multilateral arrangements are usually foreign to its elements. To the extent that overwhelming predation pressure compelled prehistoric societies to act collectively in mutual defense to the exclusion of private initiatives, it could perhaps facilitate the evolution of "social contracts." It could hardly facilitate, however, the evolution of market exchange.

The fact that prehistoric institution of exchange evolved and to a certain degree flourished despite predation pressure is in no little part a tribute to behavioral flexibility of the type displayed, as we saw, even by the baboons; namely, their (diurnal) ability to quickly regain free interplay of individual action and interaction as soon as a collective threat to life is safely removed. Yet, the major expansions in prehistoric market activities (e.g. the introduction of money) could not have been achieved without considerable and relatively prolonged relaxation of predation pressures in certain pockets of the environment accessible to humans but, less so, to their major predators. The possible existence of such rare pockets of tranquility in time and space is explored, in juxtaposition with data from the prehistoric record, in the second part of this volume (especially in Chapter 11). In the meantime, the discussion in the present chapter further follows the natural history of the baboons with an emphasis on analogies that may have additional bearings on the natural history of our distant ancestors.

Adaptive radiation in the baboons

The geographic distribution of the greater family of baboons spans from the edge of the rain forest (where the forest baboons, the *drills* and *mandrills*, live) to semi-desert open country (the range of the *hamadryas*), and to altitudes up to 9,000 feet (home to the *geladas*). The savanna baboons that dominate the broad niche in between come in four subspecies (*anubis, chacma, papio,* and the *yellow baboon*). What is remarkable in this kaleidoscope of species and subspecies is the capability of the baboons to maintain a degree of genetic separation (enough to project adaptive radiation) despite the absence of sharp geographic divides, and despite the presence of hybrids along seamless boundaries that separate them.

This capacity for speciation can be explained in part by the fact that the savanna offers ample opportunities for species to expand into distinct, if not sharply separated, niches. The savanna, after all, is not a uniform featureless stretch of grassland, but a mosaic of vegetation and patchy terrain with fluctuating wet and dry weather (Bourliere and Hadley, 1970). Sources of food are more patchily distributed over space and time in the savanna than in tropical forests. Selective pressures exacted on a primate – an arboreal animal in terms of primary characteristics – greatly vary from one part of its new-found open range to another. However, all these opportunities for speciation count for nothing without isolation mechanisms. The isolation mechanisms displayed by the baboons are largely driven by behavior (for compelling examples to this effect, see Kummer, 1997:116-124). Indeed, the distinguishing features that set species of baboons apart are marked perhaps not so much by differences in morphology, as by differences in behavior and social organization. The baboons are in this sense a living example of what E. O. Wilson called the *evolutionary pacemaker*: the idea that behavior should change first and then structure (1975:13). New arrivals on the evolutionary scene, and all converters to new habitats, are apt to distinguish themselves first and foremost in behavior.

The "southern ape"

On the evolutionary time scale the baboons are considered newcomers to the savanna. Logging barely 2 million years since arrival, the baboons entered the scene just in time, it seems, to take over the very niches vacated by the rapidly disappearing *Australopithecus,* the "southern ape," soon to become the lineal ancestor of the genus *Homo* (of which we *sapiens* are the only extant species). The baboon, for its part, was bound to converge on the Australopithecine model, while the human lineage in its pursuit of new niches (not least, the first dispersal out of Africa) continuously diverged away from it. Consequently, all the parallels between baboons and modern humans should pale in comparison with the analogies that could have been drawn between the living baboons and the long gone Australopithecines.

Analogies in terms of behavior, of course, are no longer recoverable from the fossil record. The physical evidence that survived the ravages of time is compelling enough, though, to suggest a measure of close ecological affinity between the two families of species. The most obvious (though highly circumstantial) testimony literally goes with the territory: the fact that virtually all the Australopithecine fossil record was unearthed in the heart of what is, or until recently was, baboon land (i.e., east and south Africa). The record itself does indicate similarity in general aspects of morphology (e.g., limbs and shoulders highly adapted for the dual purpose of terrestrial locomotion and tree climbing, body size,

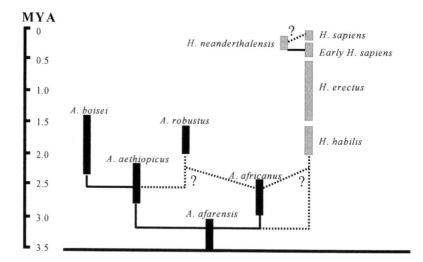

Figure 7.2 **The hominid tree of life** Schematic phylogenetic tree showing extent of adaptive radiation in the Australopithecine lineage and near lack of it in the *Homo* lineage. The general configuration of this tree is one of several alternatives under debate at the present time. Dotted lines show areas of disagreement or as yet unknown intermediate species. Depending on the species, time ranges may widely vary under different estimations and alternative interpretations.

and dimorphism) that go beyond what can be expected in a random comparison between an ape and a monkey. But perhaps the most compelling piece of evidence seems to fall under the rubric of macroevolution rather than microevolution; that is, the fact that the *Australopithecus* like the baboon maintained a multitude of species.

A proto-human bushy phylogeny is perhaps the single most important macroevolutionary feature that distinguishes the australopithecines from their human (and early human) descendants. Given the exceedingly scanty nature of the data, it is remarkable that no less than five, and probably eight, partly overlapping Australopithecine species and subspecies have already been recovered in fairly limited geographic areas in Africa. It is safe to assume that at least a few more were actually in existence. The most recent discovery of yet another specimen (Kenyanthropus) seems to lend further support to this assumption (Leakey *et al.*, 2001). In any event, it is already clear from the established fossil record (see also Figure 7.2) that the horizontal mode of evolution (cladogenesis) typical of the baboon is also applicable to *Australopithecus*; whereas, the vertical mode of evolution (anagenesis), with ancestral species being replaced sequentially in a single lineage, is applicable to human evolution only at a later stage. For additional corroborative evidence we have now to look beyond the fossil record.

Founder-effect speciation

Three million years of adaptive radiation in the savanna should indeed have left some imprints not only on the fossil record but also on the molecular record. Chief among such imprints is loss of genetic variation due to episodes of *founder-effect* speciation (Mayr, 1963; Marks, 1996). Each time a small bud from a larger ancestral population establishes itself as a newly isolated entity, the newly established species is bound to carry in its gene pool only a fraction of the genetic diversity represented in the parental population. This kind of (horizontal) speciation may be applicable, as indicated, to the Australopithecines (and the baboons), but not to the chimpanzees or the gorillas. Both the chimpanzees and the gorillas evolved largely in a vertical line of successive descent, presumably under little pressure of adaptive radiation. Indeed, as indicated earlier, they still occupy areas where minimal environmental changes have taken place since early Miocene (Lovejoy, 1981). Thus, while our nonhuman relatives were continuously accumulating genetic variation in their steady forest ecology, our early lineal ancestors were probably losing theirs under adaptive radiation in the harsher ecology of the savanna, which entails, by implication, repeated founder-effect episodes. One can thus anticipate a detectable deficit of genetic diversity in modern humans compared with apes. This theoretical expectation is nicely corroborated, it seems, by biomolecular data already in the literature for some time. These include observations (e.g., by Ferris *et al.*, 1981) detecting far less genetic diversity in mitochondrial DNA of humans than in mitochondrial DNA (of identical descent and comparable duration) of chimpanzees and gorillas – the very deficit predicted by the founder-effect, as explained above. No other explanation for this deficit, as far as I am aware, exists.

Trade and adaptive specialization

The baboon-like capacity to project adaptive radiation was lost in short order by the direct descendants of the Australopithecines. The first descendant, *H. habilis*, still showed a proclivity for speciation evident in its considerable variability in size and shape; so much so, that some authorities consider it not as a single species but, possibly, *two*. Starting with the next descendant, *H. erectus*, members of the *Homo* genus apparently would never exceed this number of species (two) at a time. In fact, for the most part the lineage was probably confined to a single species or, perhaps, a species and a single subspecies. Since the disappearance of the Neanderthals, some 30,000 years ago, a single unified human species is all but a forgone conclusion. The horizontal mode of evolution typical of the *Australopithecus* (Figure 7.2) completely gave way to the vertical mode at the latest stage of human evolution.

What is most puzzling in this turn of events is the fact that unlike *Australopithecus*, or for that matter the baboon, the human lineage was almost from inception never confined to Africa. Starting with *H. erectus*, humans spread far outside their continent of origin (Figure 11.2). More diverse niches over wider areas of the world offered them ample opportunity for (allopatric) speciation through geographic isolation. This is the opportunity that humans, with the possible exception of the

Neanderthals, never seized upon. Instead, in a course of events that requires further explanation, early humans lost their ancestral capacity for *adaptive radiation* and gained new capabilities for *adaptive specialization*. The replacement of adaptive radiation with adaptive specialization, I will now argue, was an inevitable outcome following the introduction of trade as a principle of organization in human affairs.

Wherever it goes, trade leaves no genes far behind. By its very nature, trade intensifies the frequency of communication and interaction between strangers. Reclusion and shyness are not attributes to be selected for in a species of traders – of which we humans are the sole example. By comparison, the baboons neither trade nor do they show much interest in communication with neighbors and strangers of their own species. Indeed, commenting on the frequency of communication among troop members of wild hamadryas baboons, Hans Kummer (1997:146) provides the following account:

> A baboon directs about ten times fewer gestures toward a member of its clan than toward a family member, and there is a further reduction of approximately a factor of ten in this frequency across band boundaries, so that members of different bands almost never "address" themselves to one another.

Humans do trade and, not by coincidence, do show a far greater interest in communication and free interplay with non-kin and strangers, however remote. This should lead, more readily than otherwise, to a modicum of interbreeding across barriers of genetic isolation. The precise mechanisms could operate either through geographic mobility of people who physically followed the trade routes or – perhaps more plausibly in a prehistoric setting – through gene flow.

Gene flow occurs when neighboring populations come into contact and mates are "exchanged." Just as waves of energy radiate through matter with none of the particles moving very far from their initial position, movement of genes can extend across continents with none of their (phenotypic) carriers making a move beyond their daily range. (Radiation, incidentally, is also the mechanism by which durable commodities move long distances in the manner of hand-to-hand trading typical of societies lacking specialized means of transportation (see also Box 11.2). It should come therefore as no surprise that humankind today constitutes one and the same species all over the planet. Despite (phenotypic) diversity of races, no subspecies (genotypically) exists at the present time.

Whether exchange could operate across subspecies in the same manner it operates among nonkin within our species is not clear, nor is it clear whether interbreeding human subspecies coexisted at some remote time in human history or even as recently as the last Neanderthal. It is clear, however, that under market exchange no subspecies can for any length of time maintain the degree of genetic isolation required to sustain it apart from the rootstock of the species at large. With the introduction of trade, evolution is apt therefore to shift gears, from cladogenesis (the horizontal mode) into anagenesis (the vertical mode). Now, since we know that market exchange was adopted at some point (date uncertain) in the course of human evolution, and since we also know that the process itself underwent an abrupt shift from cladogenesis to anagenesis (reason uncertain), it is not unreasonable to link the two events in time and cause.

Implicit in this logic is the striking conclusion that the antiquity of human exchange can be traced back to an early time range of 1.5–2 million years ago. This conclusion may disturb, it seems, some established conceptions about the range of capabilities we usually attribute to our early small-brained ancestors at such a remote stage in the course of human evolution. But to assume otherwise is to raise a question no less disturbing. Why would a multispecies protohuman lineage idle its propensity for speciation just in time for its first intercontinental dispersal (out of Africa) and hardly invoke it again? If anything, the dispersal to a new diverse and ever expanding geographic range riddled with formidable isolating obstacles should, in all likelihood, work in favor of speciation rather than against it. To argue that the collapse of the hominid family into a single *Homo* lineage predated and came to pass without the aid of exchange would, at the very least, require an appropriate alternative explanation. As of this writing, I am aware of no such alternative explanation. The fact that only one human species walks the earth at the present time is, consequently, apt to be poorly understood if one disregards exchange.

With this understanding, it is only prudent not to rule out the time range of 1.5–2 million years ago as a possible – if not probable – lower bound on the antiquity of human exchange. Much more investigation will be needed in order to make an unwavering determination in this matter, and some (including additional corroborating evidence) will be offered in the course of the following discussion in this volume. In the meantime, one point which is more firmly established by the forgoing conclusion is that the antiquity of exchange cannot conceivably be pushed back far beyond 2 million years, or any other date that better marks the dawn of the *Homo* lineage. If only by the virtue of their capacity for adaptive radiation, we cannot blame the inception of trade on our multi-species Australopithecine forerunners.

Part 2
Paleoeconomics

8 Departure from the feed-as-you-go strategy

The first appearance of manufactured stone tools in widespread systematic human use, about 2 million years ago, in combination with other evidence (e.g., food sharing), suggests the possibility that some form of exchange, however primitive, was already practiced by traders with brains half the size of a modern human being. If true, the implication is that exchange could have played an important facilitating role already in the first major economic transition in human (or protohuman) history – the transition from the feed-as-you-go strategy of primate feeding to hunting-gathering. The discussion in this chapter will consider these possibilities in light of the available paleoarcheological data, starting with a preliminary review of the physical environment.

The physical environment

The single most important factor that describes the theater of events in evolution, certainly in human evolution, is climate. The global climate in which civilization flourished is the wrong environmental model for understanding human evolution not only throughout the long ice ages (nearly 90% of the time humans walk the earth) but, in all likelihood, also in the course of the short spikes of interglacial periods (the remaining 10%). The climate observed today and enjoyed by (anatomically) modern people over the past 10,000 years is a placidly warm and, arguably, all too brief interval in a long process of falling temperatures starting about 3.2 million years ago (and greatly intensifying roughly 2.4 million years ago and then again about 900,000 years ago).

Any attempt to impose the template of the current environment on early humans is liable to miss the key feature of the ancestral milieu – the glacial experience. For the better part of their tenure, humans and protohumans evolved on the surface of a cooler, drier and, by all likelihoods, a far stormier planet than ours. More often than not over the last million years the earth has been locked in the deep cold of ice ages that lingered on for 100,000 years or so at a time and were interrupted only by brief interglacial intervals: spikes of mild weather as short as 10,000 years, the most recent of which (i.e., the *Holocene*) is now enjoyed by us. Much of Europe and North America – including London, New York, Chicago, and everything to the north – were buried during the frigid depths of these rhythmic glaciations under great sheets of ice thicker than Big Ben, the Empire State Building, and the Sears Tower put one on top of the other. In fact, at their apex, the ice sheets over the continents mushroomed to as high as 6,500 feet (2,000 meters), depressing under their immense weight the surface below and at certain points causing the earth's crust to sink nearly a kilometer into the underlying mantle. To this very day, the surface rocks of central Greenland are depressed almost to sea level under the weight of the ice cap. Elsewhere, as the ice retreated during interglaciation, the surface has slowly rebounded and probably keeps rising. Certain parts of eastern Sweden and the western coast of Finland are now some 100 meters above their level in the last glacial maximum. Evidently, the earth's mantle is a fairly plastic medium, by geological standards.

By their sheer mass, the ice caps on land soaked up so much of the world's water that global sea levels dropped by hundreds of feet (130–160 meters). In some spots the coast was exposed hundreds of miles from where it is now, connecting previously (and presently) isolated land-masses, and making them accessible to human and animal migration either by dry land bridges (e.g., from Siberia to Alaska across the Bering Strait), or by very short journeys over water (e.g., from mainland Asia to Australia through Indonesia and New Guinea). At the same time, and not independently, about half the planet was occupied by deserts – hot ones between the tropics and cold ones in the upper latitudes – for much of the water was locked up in the ice sheets and thus unavailable to fall as precipitation. Timberlines were drawn stringently below their present expansion trimming off so much woodland that the Amazon rain forest was on the verge of becoming a desert. Based on substantial evidence from new data (to be discussed in Chapter 12) it has recently become clear that glaciation was more than simply a matter of long-term chang-ing trends in global temperature and precipitation to which living things could comfortably adapt by incremental adjustments, as former-ly believed. It was, instead, a period of short and sudden swings of climate, of double-digit global temperature changes that may have taken place over decades, rather than millennia. The ice age atmosphere was, above all, far stormier than the atmosphere experienced by humans over the last 10,000 years. Climate volatility on such a grand scale can be, perhaps, imagined only by modern people who have had first-hand experience with the devastating effect of El Niño (a severe climate disruption to wide regions in South America and around the Pacific Ocean occurring, periodically, every five to eight years).

Just as there is risk of oversight – leaving harsh ancient climates unaccounted for – so too there is the risk of overplaying the environmental argument. A harsh environment is an absolute disadvantage to the individual and perhaps a population, but not necessarily to the species viewed as a unit of selection. In the calculus of interspecific selection, as distinct from intraspecific selection, one has to factor in the impact of the same harsh environment on the survival of other species – harmful to the species in question, or beneficial to it. The tendency to empathize with our ancestors who suffered through the hardship of glaciation is understandable, but the question that really matters to the evolution of the human species is the comparative hardship imposed on the competition: predators, parasites, pests, and weeds. This includes, in the first approximation, the large carnivores (e.g., lions, leopards, cheetahs, hyenas, and their like) that prey on or scavenge hoofed and other grazing animals, thus competing for a major resource in human prehistoric subsistence – to say nothing of the risk they pose to human life itself. The fact that at the present time the distribution of such big cats and hyenas rarely extends beyond the tropics and subtropics (the main exceptions are medium size and small cats such as the snow leopard and the lynx) indicates that glaciation must have acted to roll back their geographic distribution, or at least reduce their presence in the upper latitude, to the great benefit of human populations to the north. I will be more explicit about this point in a subsequent chapter.

The major brunt of the ice age was undoubtedly borne not by mammals but by insects and reptiles who are poorly equipped to cope with the cold. The fact that the concentration of methane in the atmosphere during the most recent glaciation was well below its present and previous interglacial levels (see Figure 12.1) indicates that amphibians and other forms of life that rely on wetlands could not fare much better.

Swamps and marshes, the main source of methane, probably dried out or froze. Yet, neither reptiles nor amphibians, nor even insects, have ever loomed large in human subsistence. The two major products harvested from insects (honey and silk) have ancient origins but only secondary economic impact. It is true that the existence of insects, reptiles, and amphibians is indispensable to the integrity of the larger food chain. Thus, for instance, many species of birds depend on insects for food as do flowering plants, for pollination. Their absence, or even their seasonal disappearance, may alter the landscape of fauna and flora. It seems, however, that the meager representation of these small creatures in the upper latitudes during glaciation – as is true to a lesser extent even today – did not reach a point detrimental to human existence.

The ice age may have actually worked to the net advantage of humans (as a species). Deforestation increased the range of grazing game, and falling temperatures reduced the presence of large carnivores that prey on such game, and on humans. Glaciation may have even helped to create additional resources: stores of durable protein in the form of frozen carcasses preserved under tundra or permafrost uncontested condition, for once frozen, the meat is no longer edible to nonhuman scavengers. The control of fire would have allowed humans to defrost and readily use such resources (Stringer and Gamble, 1993). Indeed, fire, protective clothing, and construction of dwellings gave humans a clear edge (see Chapter 11). For better or worse, the only serious complaint that humans (as a species) can make against glaciation is that it delayed for far too long the arrival of agriculture (see Chapter 12).

Stone tool technology according to Darwin

As already indicated, to my knowledge Darwin's silence concerning the matter of exchange – as an economic issue in human evolution – was broken only once, indirectly, and in the most laconic of comments. In a passage that holds three important ideas, he expressed his position regarding stone tools ascribing the observed refinements in this ancient technology, in large part, to division of labor and the institution of exchange. The implication, I think, is that one can infer from observed refinements in stone tools that occur in the archeological record about the volume and intensity of exchange in ancient populations – the makers and users of these tools. I will make use of this approach rather extensively in later parts of this book (especially in relation to the Middle to Upper Paleolithic transition, in Chapter 11).

Responding to Schoolcraft (apparently, a contemporary archeologist), this is what Darwin had to say about the subject:

> Even to hammer with precision is no easy matter, as every one who has tried to learn carpentry will admit. . . . To chip a flint into the rudest tool, or to form a barbed spear or hook from a bone, demands the use of a perfect hand; for, as a most capable judge, Mr. Schoolcraft, remarks, the shaping fragments of stone into knives, lances, or arrow-heads, shows "extraordinary ability and long practice." *This is to a great extent proved by the fact that primeval men practised a division of labour; each man did not manufacture his own flint tools or rude pottery, but certain individuals appear to have devoted themselves to such work, no doubt receiving in exchange the produce of the chase.* (1874:49, emphasis added)

As expected, an adaptively perfected organ – the human hand, in this case – is to Darwin the paramount part of the explanation. But to achieve "extraordinary ability" through "long practice" in the lifetime of an early stone tool maker is a feat that seems to demand further explanation. Darwin chooses to bet on a secondary explanation that reads like a line freshly replayed from the *Wealth of Nations* – complete with exchange and division of labor (as italicized, above). The only point Adam Smith could add, I surmise, is that the refinements of all tools (including perhaps the hand itself) are determined by the extent of the market.

Tools contain information not only about the objects upon which they were designed to operate, but also about the person who designed and used them. The user's day-to-day occupation and major source of subsistence are often revealed in the tools he or she carried or left behind in the archeological record. At the level of society as a whole, tools – or specialized organs if the society is composed of insects – are testimony to the nature and the level of interaction among its members. Purely from an economic point of view, tools are first and foremost a measure of division of labor and, by extension, of exchange. Inferences drawn from tools and implements are fairly reliable, at least as they bear upon economic structures and strategies. On the other hand, attempts to draw similar inferences about the higher faculties of the human mind – the level of intelligence and cultural progress – are far more problematic. There is little proof that modern treatises in philosophy or great works of drama written with the aid of powerful word processors are all superior to the handwritten works of Aristotle or Shakespeare. Mathematicians in their most creative work almost invariably keep using little more than a pad and a pencil, proving thousands of new theorems per year – and the concept of the computer was theirs to begin with. It is true that advances in philosophy, drama, and mathematics are endeavors that hardly rely on implements and utensils, whereas other human endeavors do. Still, it is my strong suspicion that we can only look in vain for a correlation between the degree of sophistication of tools and that of their users not only in a single generation, but also throughout the entire span of anatomically modern humans.

The real price of a tool, roughly speaking, is its cost in acquisition (or processing) divided by its time in use. The most expensive tool is thus least likely to sit idle. The most highly refined and specialized, and thus expensive, tool is likely to be picked out by a user who has the most frequent need to put it to use, and not necessarily the user who scores highest on the IQ test. The unspecialized user is well advised to acquire an all-purpose tool, and the same advice goes for a user on the road and all nomads who wish to rid themselves of excess baggage. Can we apply this logic to pre-modern humans?

The striking correlation between a brain that grew larger and larger and the use of increasingly refined manufactured tools may lead to a different conclusion. It is this correlation, more than any other thing that meets the eye, that probably led archeologists to view tools – especially stone tools – as a measure of prehistoric human intelligence, although they would generally agree that no such correlation is evident in more recent historical times or in current affairs. Economists who deal primarily with current affairs attribute, as pointed out, refinements in the manufacturing and use of tools to the degree of division of labor and specialization in society and, by implication, to the extent of the market and the intensity of exchange.

There is no fundamental contradiction between the two approaches. The inconsistency, if any, is strictly in the pacemakers used to simulate and regulate the inquiry in different fields of study. So long as the pacer is properly set to tick in evolutionary time (i.e., when the basic units of observation are successive species or populations spaced many thousands of generations apart, and consequently highly variable in brain size) it is safe to assume that any observed refinement in tools is nearly always matched by a corresponding refinement in cognitive capacities. Economists have to contend, however, with phenomena that rarely exceed the current generation. At this level of organization, they are perfectly justified to take the brain as constant and ascribe any observed refinement in tools to other agencies.

There are intermediate cases. For instance, the Middle to Upper Paleolithic transition (to be discussed in Chapter 11) was roughly 10,000 years in the making: far too long for an economist and slightly short for a student of evolution. Which pacemaker was ticking during that time? Was Darwin right to make a connection between flint tools and exchange? These are some of the questions to which we seek answers in the following discussions.

Exchange augmented food-sharing

If we could interview a chimpanzee about the behavioral differences separating us from them, what item would it find most impressive? The late Glynn Isaac who first raised this question was one of the most eminent paleoarcheologists of the twentieth century, and a leading authority on the hominid lifestyle and evolution in the time range 1.5–2 million years ago: a period of great importance to the present discussion. As to the response on the part of the chimpanzee, Isaac surmised it might go along the following lines: "These humans get food and instead of eating it promptly like any sensible ape, they haul it off and share it with others" (Isaac, 1983). The chimpanzee relies on a feed-as-you-go system of subsistence, as do all apes and most other members of the primate family, and as such, does not excel in the arts of food-sharing and food-transport. Humans are an exception in this respect. The first appearance of a proclivity for food-sharing in the hominid lineage is therefore an event of special interest to the study of human evolution.

The "food-sharing hypothesis" developed by Isaac and his associates was an attempt to provide an overall interpretation for the existence of concentrated patches of discarded human artifacts heavily associated with animal remains in layers between 2 and 1.5 million years ago in several early hominid East African sites. Finished stone tools along with raw material from distant sources of flakeable rocks (10 km or more away) occur in these sites in close juxtaposition with dense clusters of broken-up bones bearing cutmarks inflicted by stone flakes (and sometimes overlapping, in different configurations, with tooth marks left by carnivores or scavengers). These findings most directly suggest the use of tools and meat eating. Indirectly, they seem to suggest radial ranging patterns of food acquisition, postponement of consumption, and transport of some foodstuffs back to a home base or a central site (where the concentrated patches of bones and tools were actually found). At a much higher level of conjecture, the findings may also suggest the possibility of division of labor and collective consumption (especially of meat) "by members of a social group, some of whom, especially females and young, had not participated in its acquisition" (ibid., Isaac, p. 535). This concept of communal consumption apparently served as a source from which the term "food sharing" was initially derived.

Unfortunately, the term "food sharing" probably helped draw attention away from the main point of Glynn Isaac's observations. Once it enters wide circulation, this term tends to evoke an unintended "heart-warming" feeling that may obscure the issue.[1] Aware of the problem, Isaac tried to distance himself from this unhelpful modicum of "warm glow," if not from the term itself, by emphasizing that conscious motivation for "sharing" need not have been involved:

> The food-sharing model has been widely misunderstood as implying . . . friendly, cuddly, cooperative human-like hominids. This need not be so. The attractiveness of this model is that it seems entirely feasible for such a behavioural system to come into existence among non-human hominids that had brains no larger than those of living apes, and it is my strong suspicion that if we had these hominids alive today, we would have to put them in zoos, not in academies. (Ibid.)

---

[1] The existence of this sensation (in nepotistic settings) or lack of it (in market settings) is in itself a subtle issue in human evolution that, as we already saw, Adam Smith was able to keenly identify for us (see the discussion in Chapter 2).

Facing mounting criticism on the grounds of natural history (Lovejoy, 1981) and on the point of archeological methodology (Binford, 1981, 1985), as well as new evidence and alternative interpretations (Potts and Shipman, 1981; Shipman, 1986; and Potts, 1988) he slightly retracted his position suggesting that "The food-sharing hypothesis should be renamed the central-place foraging hypothesis" (*ibid.*). In my opinion, the entire issue revolves around a question that has far-reaching implications in economics: *redistribution.* I don't know if Isaac would approve of this restrictive label. The fact remains, however, that his contributions are closely associated with two important facets of redistribution, as the practice evolved in nature and in the course of early human history.

Redistribution occurs whenever a scarce resource is made available to individuals who took no part in its initial acquisition or processing. Redistribution of food in the context of kinship and reproduction is a fairly widespread phenomenon especially in the feeding ecology of animals that care for their young. Beyond parental care, food sharing is occasionally observed in courtship rituals, pair formation, and group bonding. Central-place foraging, including food transport – a special manifestation of this phenomenon – is most frequently observed in nesting animals (especially birds and insects), and occasionally in free-ranging animals (e.g., wild dogs), but with the exception of humans, rarely, if ever, among primates. The first appearance in the human record of central-place foraging (as detected by Isaac and his associates) is not only a major human departure from the feed-as-you-go ecology of primates, but may also signify a more fundamental human departure from the feeding strategy of animals at large.

Quite apart from the question of provisioning via a central place is the matter of redistribution in large groups where the stringent barriers of kinship are no longer well defined. From an archeological point of view this matter is perhaps the most speculative aspect of the entire food-sharing (or central-place) hypothesis. But it is also, at least from an economic viewpoint, the most intriguing aspect. If the hypothesis is proven to be true in this particular sense, then market exchange implications would follow almost as a foregone conclusion.

Isaac originally envisaged large social aggregations engaged in "collective acquisition" of food and "communal consumption" as part of his food-sharing/central-place foraging hypothesis. Given the sheer size of the bone/stone accumulation in some of the sites (e.g., more than 40,000 bone fragments and more than 2,600 stone artifacts in FLK Zinj site in Olduvai Bed I alone), it is not unreasonable to conclude that despite the early date of 1.8 million years ago, the original dwellers occupied these sites in numbers that far exceeded the size of a single kinship group.[2] Though Isaac's hypothesis was seriously challenged on many other issues by a number of alternative archeological interpretations (as already mentioned, but especially, Binford, 1981), little attention was paid to the conditions under which food-sharing or "communal consumption" is a viable redistribution system in large social aggregates (in the sense that it can support, and be supported, by a sustainable all-encompassing subsistence system). The lessons we learn from both natural history and economic history leave very few options in this regard. All organisms that rely on sexual reproduction have only two known options (others, notably colonial animals that reproduce asexually, have more). The first option is *eusociality* – an adaptation that evolved to a high degree of perfection (as noted in Chapter 4) in the social insects and, to a lesser degree, in certain species of mammals and at least one species of marine invertebrate. *Eusociality* entails, however, a lifestyle in which breeding is primarily limited to a single reproductive female functioning as a queen in a society of permanent or temporary sterile workers. The only other option is market exchange – observed only in humans and proved to be perhaps the only workable option open to them. The attempt to replace the marketplace with a more benevolent institution seems so far to have produced, for better or worse, only ill-fated results. Hence, one way to make (economic) sense of the food-sharing/central-place foraging hypothesis is to view it in the context of some early form of market exchange – however primitive by modern human standards. The only alternative, it seems, is to contemplate an equally early "eusocial phase" in human evolution. While both options are possible, one must agree that the primary surviving characteristics that endured up to modern humans, and are evident in our morphology and behavior, strongly favor exchange.

Further support for the possible existence of some rudimentary forms of exchange at these remote times is provided by large accumulations of

---

[2] Detailed description and evaluations of the Oldowan sites in general, and the FLK Zinj excavations in particular, are provided in Leakey (1975), Bunn (1986), and in Bunn and Kroll (1986), among others.

unprocessed stones hauled from distant sources of flakeable rocks (10 kilometers or more afar) and found in close proximity to the Oldowan sites. Such caches may represent raw material depots serving local stone-tool manufacturing, or butchering and meat processing sites where the tools themselves were in heavy demand on short notice (Potts, 1988). In any event, their existence can hardly be explained (on economic grounds) in the absence of division of labor and exchange. Without a specialized class of stone-tool makers, there would be no reason to transport (on foot) heavy loads of raw material over long distances – one could self-process the tools at the source and thus spare the deadweight in transport. In the absence of exchange, unworked stones might still be hauled for further processing by more qualified family members or other close relatives, but if this is the case, there is no reason why the carriers would choose to deposit their cargo in a common stockpile.

On the other hand, in the presence of exchange, the formation of such depots can come into existence under a number of reasonable scenarios. Under exchange, transport (broadly understood) is always fueled by arbitrage: the incentive of carriers to ferry a commodity simply because its exchange value at destination exceeds the value it commands at origin. Arbitrageurs (also known as speculators) need not be aware of the fact that their action alleviates shortages, enables the efficient utilization of scarce resources, and helps to lessen or eliminate price discrepancies – nor need they necessarily care about such formidable issues any more than other members of the community. As to arbitrageurs in a paleolithic setting, it is not hard to imagine a hunter confronting some interesting stone fragments that could be exchanged back home for food. If out of luck with game on a certain day, such a hunter is not apt to return home empty handed. The least lucky of hunters or diligent gatherers in society may even choose stone hauling as a way of making a living. Arbitrage in tangible commodities such as raw materials and foodstuff (or stonestuff in a lithic economy) is probably the most transparent rendition of redistribution in any exchange economy, simply because large visible objects move far from their source. For instance, far late in the Paleolithic age, when exchange was already a firmly established practice (but means of transportation were still confined to human carriers moving on foot), particularly desirable types of flint were routinely transported over the great plain of north-central Europe to distances as far as

100–200 km or more from their source (Schild, 1984). The fact that unprocessed stones were hauled to the Oldowan sites 10 km from their source should come as no surprise assuming, of course, that some form of exchange was already in place by the earliest of stone ages.

There are more subtle renditions of the practice, chief among them is arbitrage in the temporal (as distinct from the spatial) dimension, where agents hedge against the future by stockpiling in the present and maintaining inventories of specific commodities over time. Any society that operates under the rules of market exchange, if it is to be suddenly wiped out, is bound therefore to leave in the record "unexplained" accumulations of certain distinct commodities and, on the whole, a far more patchy scatter of artifacts than a society that operates without the benefits of exchange. Such accumulations are occasionally found in archeological sites – with the Oldowan assemblages of unprocessed stones being the earliest but by no means the last.

Could market exchange in some form, however rudimentary, have come into existence among early humans or prehuman hominids some 1.8 million years ago? Any suggestion to this effect is, I admit again, so much at odds with existing theoretical expectations that it pushes the envelope of antiquity to its limit, if not beyond. To draw such a suspect conclusion from a single reasoned argument is always a risky exercise. But by now we have two arguments derived from separate sets of logic and supported by two independent bodies of evidence: the argument from macroevolutionary considerations (outlined in Chapter 7), and the argument from the food-sharing/central-place foraging hypothesis outlined above. The upshot is that both arguments point in the same direction favoring the early inception of exchange in roughly the same spatial and temporal setting, one that "happens" to coincide with a cardinal event in human evolution – the very dawn of the *Homo* genus.

Two distinguishing features mark the inception of the Homo genus: a large brain and the extensive use of manufactured tools. *Homo habilis*, the founding father of the genus, is hardly distinguishable from its Australopithecine contemporaries in any major morphological feature other than an enlarged braincase (volume of about 700 ml compared with 500 ml). If not for the massive accumulations of stone artifacts and debris typical of an ardent tool maker and user, *H. habilis* would run the risk of passing unnoticed in the archeological record. The observed early correlation between a large and growing brain and the extensive use of increasingly refined tools also carried over to later periods, and came to dominate the entire inquiry into the evolution of humans throughout the ensuing stone ages. The correlation is occasionally extrapolated even into the tenure of anatomically modern humans. For instance, one of the leading explanations for the Upper Paleolithic "creative explosion" that took place as recently as 40,000 to 30,000 years ago, and involved a sudden refinement in material structures and behavior without an apparent change in human morphology, ascribes the phenomenon to some unobservable neurological change (Klein, 1989:409–410). A sudden intensification in exchange under certain plausible conditions (to be discussed in Chapter 11) fits neatly into this gap at least as a possible explanation.

In sum, Isaac inferred from the data at his disposal indications for division of labor (with respect to food procurement) in combination with food sharing and interpreted it as a pivotal adaptation in human evolution. But the adaptive mechanism that regulates division of labor and food sharing (i.e., the distribution system) which was so essential to the argument, was not spelled out explicitly. Specifically, the possibility that exchange could provide such a mechanism has not been fully appreciated, although, with exchange in place, division of labor and food sharing are both attained as a foregone conclusion. Of course, doubts about the possibility that a hominid with a brain half the volume of a modern human (though already one third larger than of a similar sized ape) was engaged in some form of exchange some 1.8 million years ago are well taken. On the other hand, in light of the discussion in Chapter 5, it is also hard to see how such a half-sized brain could stand a chance of doubling without its carrier being engaged in some form of exchange, to begin with.

9   The origins of market exchange

If market exchange evolved from some preadaptation observable in humans or animals, then a careful examination of analogies with nepotistic exchange (such as courtship feeding in birds) or symbiotic exchange (such as pollination) is a logical course of action to pursue. In following this approach the first part of this chapter demonstrates the existence of some tempting analogies of market exchange drawn from the behavior of certain animals and plants, as well as from the nonmarket sphere of human affairs. It also demonstrates the risk of drawing premature, if not fanciful, explanations from such analogies.

Alternatively, if market exchange evolved *de novo*, then the major thrust in the inquiry is best directed toward the mechanisms of the market and the deep structures of exchange itself. This approach is undertaken in the second part of this chapter with the hope that intricacies of market exchange as we know them could provide a clue to their origin. At issue is a catalyst in the form of an activity, or perhaps a single commodity, with the power of spurring exchange between exceedingly reluctant traders at some remote point in antiquity.

Bateman's syndrome

There are certain parallels between the choice of mating partners and the choice of trading partners – chief among them is asymmetry. The asymmetry in the case of sexual selection is due to adaptive pressures that for a widely recognized reason – known as *Bateman's principle* – act differently on the reproductive behavior of males and females. In a pattern that cuts across nearly all the species of higher organisms, as already noted by Darwin (1874), females are more selective. Males, on their part, exert far less choice and far more effort in an attempt to (1) drive away or eliminate the competition from rivals of their own sex, and (2) gain access (e.g., through display and courtship) to discriminating mates of the opposite sex. The explanation by Bateman (1948) was based on the difference in the gametic contribution made by each parent to the offspring. The female investment in each fertilized egg exceeds by far the male investment in the fertilizing sperm. For that reason, eggs are produced in numbers and, more importantly, frequencies far smaller than sperms. The degree of reproductive success of females is thus largely independent of the number of mates they have in each reproductive cycle. In contrast, the degree of reproductive success of males is closely related to number of mates they have. Females, the providers of the limiting resource in this gambit, are consequently an object of *male competition*; and males, an object of *female choice*. This argument was reinforced by Trivers (1972) by extending the notion of investment to include parental care.

A corresponding asymmetry exists in the marketplace. It is evident in the relationship between customers and vendors, employers and employees, professionals and clients, and between principles and agents, in general. Consider, for instance, the retail industry. In close analogy with the female role in sexual selection, customers (of both sexes) seem to be highly inquisitive about the merchandise they seek to acquire; thus collecting information, checking warranties, sorting for quality, trying for size and color – always comparing prices with alternative retail outlets in search of a better bargain or better service. In short, they are primarily engaged in search and the exertion of choice, for they bear the cost of the final price. Vendors, on their part, act in analogy closer to the role of the male. First they try to eliminate the competition from their own ranks, sometimes to the point of cut-throat price wars. Then, they spend a fortune on display and advertisement. In sum, the retail business shows signs syndromic to Bateman's principle. Can we trace the origin of market exchange to sexual selection?

Obviously, such a conclusion seems to be a convenient detour and a tempting answer to an important question. But it will be the wrong detour (and quite probably the wrong answer). First, there is the risk of stretching an analogy beyond what it stands for; namely, a similarity in one respect between unlike things that are otherwise not comparable. Sexual selection, as is true of kin selection and nepotistic activities in general, is comparable to market activities in none of its more fundamental functions (see discussion in Chapter 2). In this sense the analogy is quite shallow. It would be a logically misguiding error to infer from the fact that the two activities share a common feature, Bateman's syndrome, that they must for that reason alone be identical in any other respects, let alone the possibility that one is a derivative of the other. In fact, it is quite likely that Bateman's syndrome is a run-of-the-mill phenomenon that repeats itself time and again in many other unrelated situations. To be sure, let us go completely outside the spheres of either market or nepotistic activities to a purely symbiotic relationship between a plant and an animal: pollination.

Bees, the most important insect pollinators, play the role of the "choosey" (analogous, above, to "female" or "customer") partner in their relationship with flowering plants. They are heavily engaged in search and detection – collecting, sorting, and transmitting information. To that end, they possess well-designed adaptations. Their sense of vision ranges into the ultraviolet wavelength of light (not visible to the human eye) and they are keenly sensitive to odors. They have a refined sense of sweet tasting that discriminates among different kinds of sugars, which comes as no surprise for a harvester of nectar. Honeybees are best known, of course, for their ability to communicate to one another both the distance and the direction of food sources through dance. The exertion of choice is best demonstrated by the fact that the flowers visited by the bee are not chosen at random. Honeybees will prefer, for instance, to stay with one species of flower – typically the most abundant – to the exclusion of all others. Economists call it weak "brand loyalty" combined with strong "networks effect" (see Box 9.1). Flowering plants respond by playing the role of the vendor (or the "male") in their relationship with their potential pollinators. They use a display of bright colors to attract the insects from a distance, and special fragrances at close range. In close analogy with a parking lot in front of a suburban store, certain flowers even provide their pollinating patrons with a special landing platform on which a visiting insect can recuperate before it penetrates the flower's depth. (This function is served by the broad lower lip visible in some wild flowers but no longer in most cultivated ones.) The story of pollination complete with Bateman's syndrome is replayed with variation in other symbiotic relationships, not least those involving a human player. Can we trace the origin of market exchange to symbiosis?

Once again, standing alone such a conclusion may seem as tempting (or as absurd) as the one suggested earlier with respect to sexual selection. Taken together, they rule each other out and more nearly suggest

Box 9.1 Macintosh: the computer and the apple tree
To the near demise of Apple Computer Inc., the company failed to realize in the early 1990s that its top-of-the-line (and industry forerunner) product, the Macintosh computer, faces the challenge of attracting customers not far different from the challenge facing the original Macintosh tree in attracting its own sort of customers: insect pollinators. Honeybees, among other pollinators, can pollinate many species of plants but nearly always choose to stay with only one at a time, largely depending on the abundance of flowers (a crucial coevolutionary adaptation, since pollination is effective only when pollen is transferred between plants of the same species). Computer users can handle many brands but prefer to stay with one brand – typically the most abundant in the supply of compatible software. Economists, in their own language, delineate the situation by the phrase: weak "brand loyalty" strong "networks effect." Ironically, the situation Apple Computer faced but apparently failed to recognize was long known in the business world. A similar situation can be expected in the market for nearly every durable appliance requiring ongoing supply of some peripheral media. Examples go back at least to the oil lamp (dependent on the supply of kerosene) and the invention by Gillette of the safety razor (likewise dependent on the supply of fresh compatible blades). In the heyday of kerosene lighting the former was distributed free of charge by Standard Oil Co. in China, and to this day the latter is available at symbolic price from many drugstores throughout the world. If one wonders why apple trees in blossom seem to be burdened with flowers to the point of overabundance, think about their competition with other flowering plants under the "network effect" strategy (the very strategy used to great effect by Standard Oil and Gillette, and more recently, by Microsoft against Apple and other competitors).

that Bateman's syndrome, broadly defined, tends to repeat itself in all reciprocal interactions that require some prior act of pairwise matching between asymmetrical parties. In this sense, the principle is too generic to be of any use as a distinguishing feature that sets apart a particular set of reciprocal relationships from another. What is also clear is that

analogies either from nepotistic or symbiotic exchange, however tempting, bring us no closer to a resolution concerning the historical origins
of market behavior. The alternative, as already suggested, is to direct
more careful attention toward the deeper mechanisms and structures of
exchange itself.

The impetus to trade

A gambit, in the game of chess, is an expedient move in which a piece is
offered in exchange for a favorable position. It always opens a real or
imaginary window of vulnerability. Given the opportunity, most novices
are at first highly reluctant to take advantage of it. This common inhibition, it seems, can only be overcome by a reward of great inducement
and not without agony. The improved position must be apparent on the
current chessboard or near at hand (in the next move), and the payoff
must far exceed the actual sacrifice. In other words, the threshold of the
novice is capriciously high.

Exchange is a *reciprocal* gambit played in a different game (cooperative,
rather than zero-sum) and according to different rules, but the inhibition – unless muted by proper adaptations – is fundamentally the same.
The compulsive inhibition of entering mutually beneficial give-and-take
relationships, the essence of market exchange, is the default mode of
animal behavior. The exceptions, as we saw, are well defined and fairly
predictable: symbiotic exchange (across species) at first approximation
and nepotistic exchange (among kin and mates) at the next. Mercantile
exchange (between conspecifics at large) as practiced probably only by
humans is already an exception of the third degree. It is, understandably,
the most striking form of exchange because it takes place where the competition in the ordinary affairs of natural selection is most fierce: neither across distant species nor among close relatives, but between unrelated members of the same species.

Modern people are veterans of the game. Thanks to thousands of generations that have experienced the advantages of free trade and occasionally the dire consequences of abandoning it, we are no longer reluctant to cut (bilateral) commercial deals with each other. But early on, the
very first converts to the games of trade had to overcome the full brunt
of the inhibition – unmuted and undiminished – and without the aid of

a (genetically, let alone, culturally) preconceived notion of exchange at their disposal. The only thing that could have possibly spurred exchange between such early traders was a persistent clear-cut opportunity to gain extraordinary large mutual benefit from the activity. In terms of the impetus to trade, the threshold necessary to trigger a transaction between them far exceeded the threshold necessary to trigger a similar transaction between veterans like us. This understanding should, in a way, help us trace exchange closer to its roots simply because it narrows the range of commodities (i.e., objects of exchange) that are reasonable candidates for scrutiny. This is not to say that we can open up a merchandise catalog and point with a great degree of certainty to the specific item that started it all going in the first place. What we can do with some degree of certainty is figure out, however generically, at least the relevant *properties* of such a commodity – and then rule out the rest.

The nature of commodities and the structure of markets

Two salient properties present in any commodity, and in any object of exchange, shape the nature of the interaction between the traders. They largely predetermine the competitive structure of the market in which the commodity will be traded, the impetus of traders to partake in the exchange and, by extension, the prospect of that market coming into existence. These properties include, first, a measure of *exclusion* (essentially, the degree to which a producer can control access to the product by the final consumer) and, second, a measure of *rivalry* (essentially, the difficulty with which a consumer can share the product with others without detracting from own consumption). A schematic classification of goods by these two salient properties is outlined in Table 9.1. The four possible configurations define four distinct categories of goods, as listed in the table: (1) private goods; (2) public goods; (3) common property; and (4) contrived commodities. These categories are pertinent to a modern economy as much as to a prehistorical economy of hunter-gatherers, and in fact, they carry over to nonhuman populations. The impetus to trade in each case (formally inferred from the corresponding shaded area in Figure 9.1) is essentially the combined net benefit gained by buyers and sellers engaged in transactions that can potentially take place. The impetus to trade reaches its peak with *contrived commodities*, to which I

Table 9.1

	Rivalry	Non-rivalry
Exclusion	**Private Goods** Impetus to trade: Subdued	**Contrived Commodities** Impetus to trade: Strong
Non-exclusion	**Common Property** Impetus to trade: Nonexistent	**Public Goods** Impetus to trade: Weak

will return shortly. But first, a brief review of the other three categories is in order.

Private goods are by far the most familiar items of exchange; so much so, that the discourse concerning economic issues is often carried out to the exclusion of the other three categories. A potato is an example of a pure private good; street lights and the Pythagorean theorem, as we shall shortly see, are not – nor was fire before the invention of the match. A potato is a private good by the coincidence of the two characteristics just mentioned: rivalry and exclusion. Rivalry in consumption simply means that one person's consumption reduces by the same amount the quantity of a good available to others – a potato consumed by me is no longer available to you. Exclusion means that the provider (say, a potato farmer) can deny use of a product to any consumer, especially to those unwilling to pay for it. The hallmarks of private goods – rivalry combined with exclusion – are two words not known for their particular wide circulation in polite society. Yet among economists they are widely recognized as necessary conditions for the efficient operation of free markets. The efficiency of markets is not at issue in the present discussion. What is at issue is their existence in the first place. Private goods, as much as they promote efficiency in markets that already exist, are not ideal midwives to markets yet unborn. Private goods tend to fine-tune rather than maximize the impetus to trade. The point can be explained by the fact that a private good has alternative useful applications outside the transaction (i.e., opportunity costs to its seller). Unsold to the first buyer, a potato can still be sold (perhaps even at a better price) to the next. Alternatively, it can be used as seed or feed or, at worst, consumed by farmers themselves.

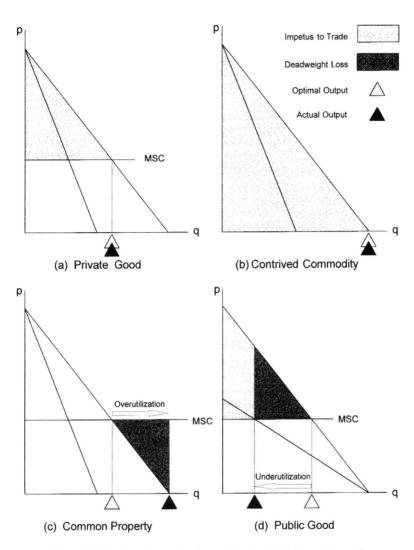

Figure 9.1 **The impetus to trade and deadweight loss** Schematic representation of the impetus to trade (lightly shaded areas) in a highly simplified (two-person) economy under four alternative market structures: (a) private good; (b) contrived commodity; (c) common property; and (d) public good. Overutilization of resources in the case of common property and underutilization of resources in the case of a public good result in deadweight losses (heavily shaded areas). The two downward sloping lines represent aggregate demand, and pro-rata demand in each case, respectively. Marginal social costs (MSC) are zero for the contrived commodity and are held constant in all other cases.

Private goods, in short, foment a modicum of prudent reluctance on the part of sellers and, as such, are not the most interesting category of commodities in our attempt to reconstruct the model of the market at the onset of human exchange.

Public goods rest on the pole opposite from private goods. Street lights and the Pythagorean theorem are public goods by virtue of being subject neither to rivalry nor to exclusion (Table 9.1). Nonrival (or joint) consumption means that one person's consumption of the good does not reduce its availability to anyone else. Thus, a public good provided to one person can be made available to others at no extra cost. Nonexclusion means that, once the good is provided, the producer can prevent no one from consuming it. In addition to street lights and the Pythagorean theorem (akin to knowledge in general), the list of public goods that play an important role in daily life is fairly long, though few observers are aware of their presence as such. National defense, certain aspects of preventive medicine (e.g., public sanitation and mosquito control), radio and television broadcasts on the air (as distinct from cable), and the beacons of a lighthouse – all share the property of nonrival (joint) consumption and nonexclusion. In its absence, some people would be willing to acquire limited amounts of the public good either for their immediate use (e.g., erecting a lamp-post on a dark street in front of one's own house), or as a voluntary action (e.g., making a donation in support of a local public TV station). However, the provision of a public good (be it street lights or TV programing) purely through such private action always falls far short of the social optimum. The main problem is free-ridership. Once the good is obtained in small amounts even by one or a few users, it becomes available without payment to all others, and soon no one would be willing to pay for it any longer. Consequently, in a modern economy, public goods are typically provided by governments and paid for through compulsory taxation. Some economists view the provision of public goods – broadly defined – as the main justification for the existence of governments. The historical background of the problem is well illustrated by the public (and sometimes private) provision of fire prevention and firefighting going back to the ancient world (see Box 9.2). Anyway, in the absence of government, private providers of public goods will have little or no incentive to operate. The impetus to trade is thus minimal if not negligible and certainly falls far behind of what is expected even of private goods.

Box 9.2 Firefighting: the private provision of a public good in ancient Rome

Firefighting throughout the Roman empire, and especially in its capital, is relatively well documented. By the second century BC there were already fire-preventing city regulations in Rome stipulating, among other things, maximum height of buildings and minimum spacing between them. During the time of the Republic, Rome had well-organized public firefighting brigades (*familia publica*) operating side by side with equally well-organized firefighting private enterprises (*familia privata*). Both forces where made up, at that time, primarily of slaves under the command of freemen and kept among them thousands of individual firefighters on duty (for a city less than a million strong).

The existence of private firefighting agencies poses a problem for economists who are always worried about the lack of adequate private incentives in the provision of public goods. Lack of incentives, however, was no problem for Marcus Licinius Crassus (115–153 BC). A Roman plutocrat and general known for his dual talent of using wealth to win office and vice versa, Crassus caught the attention of the Greek historian Plutarch and was immortalized in one of his biographies. Here, our main interest in Crassus is confined however only to his early career: an unsavory venture he pursued in the private business of firefighting. In this capacity, we are told, he assembled a selected group of 500 slaves composed primarily of expert builders and architects and forged them into a well-trained and well-equipped private team of firefighters under his command. In the event of a fire anywhere in Rome, he would rush to the scene with his men and all their equipments, but without the slightest intention of putting them immediately into action. Instead, with his team idly standing by, he was already hard at work making bids, at petty prices, for the properties under fire and those under imminent danger or uncertainty in the affected neighborhood. No homeowner in close vicinity could for long afford resisting such offers, for Crassus' bids were progressively diminishing in sight of more and more houses burning to the ground. Only when the owners gave in did he order his team to put an end to the fire in or near the properties that came into his possession. This enabled Crassus to resale or let at great profit the surviving houses to the highest bidder, sometimes even to their original owners. Between fires, Crassus' team was probably kept busy rebuilding the destroyed houses with the idea of putting them on the market at the earliest date, again, at great profit. Through this practice, Crassus came to be one of the wealthiest men in Rome at a relatively young age. "The greatest part of Rome, at one time or other" writes Plutarch "came into his hands" (1979:651).

Plutarch is known to tread his historical material first and foremost as a moralist interested not so much in political or social consequences of action undertaken by major historical figures as in their personal character (that of Crassus he seemed to despise). As such, his biographies provided excellent material for Shakespeare's plays (e.g., *Julius Caesar and Antony and Cleopatra*) but not necessarily for an overall somber evaluation of statecraft in the ancient world. Thus, the despicable character of Crassus notwithstanding, an economist cannot leave the story without asking a final irritating question: other things being equal, would the Roman public been much better off without the younger Marcus Crassus and his likes? Think about the alternative.

Crassus' attempt of privatizing a public good (by forced exclusion) was not unique in the history of firefighting. Milder and more genteel versions of the practice – coupling fire protection with fire insurance rather than with real-estate speculation – where in effect in certain European cities until fairly into recent times. Thus, for instance, following the Great Fire of London in 1666 and until about 1865 (when the Metropolitan Fire Brigade was established) fire protection in the city was relegated to private insurance companies providing the service primarily, if not exclusively, only to proprietors taking out fire insurance in advance.

An even less likely catalyst in early exchange falls under the category of *Common Property*. Common property refers essentially to a pool of scarce resources to which access is not restricted or, otherwise, a free-for-all scarce good or service. As such, it is a hybrid of a public good and a private good – sharing the central characteristic of nonexclusion with the former, and the central characteristic of rivalry with the latter (Table 9.1). The situation is sometimes known as "Tragedy of the Common" – a term popularized by G. Hardin (1968). An example of a common property resource is an open access fishery. In a lake to which an unlimited number of fishermen have access (courtesy of nonexclusion) the difficulty of catching a fish increases with the number of fishermen who actually fish in its waters (the essence of rivalry). If the number of fishermen is sufficiently large, or the lake is small, the fishery will rapidly succumb to depletion. In general, the equilibrium number of operating fishermen is bound to exceed the optimal number resulting in a twofold undesirable outcome: (1) a cut in the biomass of fishes to a level below its maximum sustainable yield and, not independently, (2) unnecessarily high harvesting cost.[1] The pattern repeats itself in similar situations associated, for instance, with game populations, grazing lands, underground water, oil fields, and congested free highways – to name a few. At equilibrium (i.e., the situation most likely to be observed), the exploitation of such open access free-for-all resources is inherently suboptimal. The proximate cause is due, of course, to overharvesting or overutilization. The ultimate cause – first stated by Ronald H. Coase (1960) and now widely accepted by economists and legal experts alike – is largely attributed to poorly defined property rights. Air, water, underground mineral deposits, and wandering populations of animals are vulnerable precisely because they all defy easy assertion of property rights. In any event, it is clear that the impetus to trade ceases completely to exist in the case of such common property resources.

So far we have ruled out three categories of commodities: private goods, public goods and common property resources. For all practical purposes, the possibility that such commodities in themselves could spur early exchange seems highly implausible. By elimination, this brings us full circle to our first category: *contrived commodities*.

Contrived commodities, as the notion is defined in Table 9.1, is another hybrid of a public good and a private good – combining exclusion with nonrival consumption. A consumer unwilling to pay for the resource can be denied access to it. But, at the same time, any additional consumer

---

[1] For a fairly complete mathematical, biological, and economic treatment of the problem, see Colin W. Clark (1976), especially Chapter 2.

willing to pay can be accommodated at no extra cost to the provider, and with no detraction in the consumption available to others. Paper money, and fiat money in general, is probably the primary example of a contrived commodity in a modern economy. Artistic lithography, taxicab medallions, copy-protected computer software, and the entire class known as *club goods* (e.g., all the empty seats in a theater or a concert hall) – are some others. Ancient systems of human subsistence had their own examples of contrived resources, as we shall shortly see. Defined by exclusion in combination with nonrival consumption, contrived resources display the reverse image of common property resources, and imply the reverse consequences. In so much as common property resources minimize the impetus to trade, contrived resources maximize it (Figure 9.1). As such, contrived commodities provide an ideal starting point in the attempt to identify and reconstruct a plausible model of the proto-market. The two entities most clearly identified as contrived commodities in the prehistoric setting of human affairs include, as far as I can see, large game and fire.

The perishable parts of large game fit the description of a contrived commodity. At least in temperate and tropical zones, hunters (or scavengers) of large game cannot on their own consume entire carcasses that come into their possession before the meat spoils. They can, however, share the meat with some members of society at no extra (opportunity) cost to themselves and, if they wish, exclude others. This produces ample opportunity for *quid pro quos* verging on exchange. As such, meat of large game quite possibly played at least a secondary role in the transition to exchange though, all things considered, not quite the primary role. This evaluation is based on two separate reasons. First, it should be noted that archaeologists generally have serious doubts about the possibility that hunting, especially large-game hunting, could have a role of any importance for subsistence in the early stages of human evolution. In addition, to the extent that humans indeed hunted large game, it should also be noted that they were hardly unique or outstanding in this respect. Small predators that hunt in groups, such as wild dogs, routinely target prey that far exceed their own size and, consequently, are predisposed to (nepotistic) food sharing – yet show no signs that nearly resemble market exchange.

Unlike large-game hunting, there are little doubts either about the antiquity of fire in human use, or about the uniqueness of this activity. Moreover, as we will shortly see, fire in its proper place and time used to be the very model of a contrived commodity. Both in its own right and by default, fire was left in a position to play the primary role in the transition to market exchange.

Fire: what's in a name?

Fire is not a commodity in the simple sense of the word. Narrowly defined, fire is a neutral chemical reaction. Broadly defined, fire is a small industry that provides occasion for a number of distinct commodities and services in all four categories discussed above. As a source of energy and in its most typical domestic applications (home cooking, heating, and lighting) or industrial applications (metallurgy, pottery, glass-making, distillation, etc.), fire was very much a *private good*. However, it should be noted that ever since fire came into widespread human use the procurement of combustive material to fuel it seemed always to be associated with overutilization of some *common property* natural resource, renewable or otherwise. All too often when a fire gets out of hand or makes one of its uninvited visits from the wild, it soon provides occasion for a *public good*: its prompt control and timely elimination. In its capacity as a public goods, firefighting (and prevention) was in fact one of the earliest services provided by ancient governments to their citizenry (but see Box 9.2 for important exceptions).

Yet the most intriguing imprint of fire on human history, at least from the viewpoint of the question at hand, was probably achieved in its capacity as a *contrived commodity* and, as such, through its incitement to trade. Fire takes the form of a contrived commodity by virtue of two peculiarities closely associated with the distinct characteristics that define such a commodity (exclusion but no rivalry). First, there is the requirement associated with all (nonspontaneous) chemical reactions: activation energy (i.e., a source of ignition). The enormity of this requirement is no longer fully appreciated by modern humans within easy reach of matches. But until a point not so distant in the past it still posed a major challenge to all fire users giving ample opportunity for exclusion. The second peculiarity of fire is its capacity for self-generation. Granted fuel, fire can propagate itself indefinitely, and its human handler can make-fire-with-fire at no extra cost. All the elements of a contrived commodity at the deliberate manipulation of humans were therefore in place not only as a prelude to history, but for several hundred thousand years before. Under the right configuration of conditions, it is not unreasonable to infer that there were many opportunities for the impetus to trade to arise and take effect. To find out more about these opportunities, and the prospects of fire as an agent of early exchange, we turn now to the story of the domestication of fire itself.

10 Domestication of fire in relation to market exchange

Calorie for calorie, fuel is less expensive than food. This is, fundamentally, the only explanation for the close tie of humans to fire – an unexpected reliance of a carbon based life form on an unruly combustive chemical reaction. The impact of fire on civilization from the hearth to the microwave oven is fairly well recorded. Yet, by comparison, the bearing of fire on human evolution prior to civilization is still poorly understood and far more speculative. Given its antiquity in human use, from 300,000 years according to fairly conservative indications, up to 1.5 million years according to others, it is hard to escape the conclusion that fire had a role to play not only as an agent of civilization but also as an agent of evolution promoting hardwired adaptations – not in the least, incitement to trade. Human use of fire, I will argue in this chapter, both facilitated, and was facilitated, by early forms of market exchange.

Nonhuman use of fire

Before we examine the use of fire by humans, it would be appropriate to examine its exploitation by certain forms of life other than human. However opportunistic and passive, the practice is apparent in plants more than in animals for a good reason. Unable to flee fire but, making a virtue of a necessity, plants developed elaborated safeguards against it and, occasionally, use its deadly reaches to their own advantage. Plants protect themselves from fire by various means (e.g., by sheltering storage and vital tissues underground) and, consequently, are fireproof to different degrees. Those which are relatively better protected often "invite" wild fires (e.g., by shedding flammable tinder) to weed out unwelcome competition, rid themselves of pests, let in more sunlight, and return minerals to the soil (see Box 10.1). Some trees, such as the sequoia, are dependent on fire for clearing the forest floor from debris that hinders new growth of their own seedlings.

The extent to which animals use fire, passively or otherwise, is more

Box 10.1 Fighting shade with forest fire, and vice versa
Consider the clash between the densely shaded rain forest and
the bright and airy Eucalyptus forest in parts of Australia where
the two ecologies share an uneasy borderline. Starting from
open woodlands or any forest clearing in the "transition zone,"
the Eucalyptus has the advantage of making the first move, simply because its seedlings best germinate and grow in bright
light, whereas seedlings of the rain forest species thrive in the
shade. However, as soon as the first vintage of Eucalyptus trees
reach maturity and their canopies grow thick enough to cast sufficient shade, some species of the rain forest that require only
partial shade to germinate get established. In time, they start to
cast their own far-deeper shade under which still other rain forest species get established. Eventually, creating its own environment, the rain forest gradually shifts the balance in its own
favor crowding out the Eucalyptus, except perhaps for some "old
growth" - surviving trees that stand out in the new environment
but can no longer reproduce in it (due to seedlings' shade mortality). The only weapon left at the disposal of the Eucalyptus is
to fight shade with (forest) fire – which creates forest clearings
in the first place – thus starting a new cycle. But how, to begin
with, can a tree make a fire and control it?

A 500-species-strong native plant of Australia (of which 35 or
so serve as home and sole source of subsistence for the koala),
the Eucalyptus starts no fires but, quite effectively, invites them
to its forest. Some species of this tree continually shed a mixture
of outer bark from their smooth greenish nearly fire-resistant
trunks along with ample foliage laced with flammable oils. The
resulting layer of debris formed on the forest floor serves as tinder ready to catch fire and blaze fiercely. The (evolutionary) strategy of the Eucalyptus tree is clearly to repel or reverse any take-
over by undergrowth. As such, it is most effective against invad-
ing tropical rank vegetation that, typically, have developed nothing nearly as effective in the way of fire resistance.

This leaves the koala with little choice but to subsist almost
exclusively on the eucalyptus. However toxic to other animals,
eucalyp leaves are the only reliable source of food and water in
the koala's flammable home range (Degabriele, 1980). The role of

> fire in the highly specialized relationship between the koala and the Eucalyptus of Australia repeats itself, it seems, in the equally specialized relationship between the giant panda and the almost equally flammable bamboo grass forests of west-central China.

questionable. There are reports of certain predators (like the cheetah in Africa) that use wild fires to ambush prey fleeing flames. Scavengers, obviously, will miss no such opportunity either. However, to most carnivores, a wild fire more nearly means a last meal than a feast. Mass destruction on the lower trophic levels of the food chain means mass starvation on the upper ones. In any event, the frequency at which wild fires occur is too slow and too irregular to be of any use to most animals. Unlike forest trees, the life spans and reproductive cycles of animals are far too short to incorporate such a sporadic phenomenon as a systematic resource in their adaptive strategy of subsistence.

Though the human claim to uniqueness in the use of fire is slightly overstated, in one fundamental sense it remains intact: humans alone use fire as a source of energy. This is accomplished primarily by manipulation of fuels.

The question of fuel

Strictly as a chemical reaction, fire relies on the supply of oxygen in combination with combustible material, to produce heat and light. Some appealing aspects of fire, from an economic point of view, are associated with the properties of oxygen. The substance is by far the most reliable resource on the face of the earth – an agent that succumbs to no scarcity. For some 600 million years it has been available in the open atmosphere at a strictly predictable and, thanks to photosynthesis, practically undiluted amount (of 21% by volume); so far, always free of charge. The fact that oxygen is a highly reactive substance by chemical standards is a virtue by economic standards. It means that there is a wide choice in fuels: wood, bone, dry dung, crop residues, charcoal, peat, coal, crude oil, natural gas – all of which have played a role replacing each other in repeated waves of substitution throughout history and across different geographic sites.

On the other hand, the acquisition of fuel to feed and maintain fire on a regular basis was never a trivial undertaking. Unlike oxygen, combustible materials are often in short supply and the most convenient fuels are rarely accessible to all users, if to any. We are all aware of energy crises, real and imaginary, in our lifetime. Historians tell us that as early as the first century BC the shortage of wood was already a major problem in certain parts of the Roman empire and, at about the same time, in Han China. The shortages, in both cases, brought forth major innovations that far increased fuel efficiency, especially in home heating. In the case of China, it even led to the use of coal in iron making. England followed in the footsteps of China shifting to coal when faced with wood shortages during the fifteenth and sixteenth centuries. The same story has repeated itself with variations many times elsewhere. In fact, in large part due to market forces, the course of history in this respect seems to fit more closely a sequence of averted, rather than actually incurred, shortages. Market forces have a tendency to eliminate any shortage before it takes place simply by converting it into a higher relative price. When conceived in a free market, a higher price is not only a symptom of hardship but also a source for the cure. It induces conservation and innovation just when they are most needed. What is special to innovations in the field of energy is that so far they have always outstripped the shortage that they came to solve; so much so, that despite a continuously increasing rate of energy consumption per capita, and despite a decreasing rate of biomass per capita, the constraint of fuel by the end of the twentieth century was less pressing and its real cost (i.e., the price net of tax) probably lower than ever before in the course of human history. Moreover, with the promise of fuel cells, if not of hydrogen fusion, this constraint stands a good chance of completely disappearing sometime not far into the new century. If this last chapter in the history of fueling defies conventional wisdom, so too did the first.

Odd as it may sound, the longest and most limiting fuel constraint was probably faced by people who enjoyed the most abundant rate of biomass per capita: the prehistoric users of fire. To appreciate their plight, it should first be noted that as a form of fuel, stemwood is far superior to brushwood and twigwood (or grass) which require frequent stocking and kindling, producing, at best, easily choked-off unstable fires. Stemwood was practically unavailable to these early users for lack of adequate logging implements (especially saws). The problem was aggravated by the lack of fire-making technologies and, hence, there was a need to feed and preserve a flame continuously. Fed only by fast burning low quality wood and dry grasses, a fire that could never be allowed to die out was bound to consume enormous amounts of fuel relative to the amount of useful energy it could possibly provide its handlers. It would soon consume all the combustible material in its near vicinity. With the ground cleared at an increasing radius around any permanent or semi-permanent site of dwelling, the provision of fuel would have to rely on longer and longer trips (by foot) and longer and longer distances of transport (on shoulder). Eventually, as the energy expended in fueling exceeded the useful energy gained from the fire, the situation would leave no choice but premature residential relocation – unless a smarter logistical solution could somehow be found. The plausibility of such a solution will be considered in a subsequent section.

Incendiary skills

The use of fire was bound to place selective pressure on the development of versatility in handling the material world in three distinct ways associated with ignition, maintenance, and containment of a largely uncontrollable chemical reaction. In the presence of (intended or unintended) combustible material and oxygen fire tends to accelerate and consume all the fuel within its reach at an ever-increasing pace and ever-rising temperature; occasionally, to the point of explosion and detonation (as typical, for instance, of "crown fires" that advance through blazing tree tops and are practically unstoppable even by the most advanced fire extinguishing equipment). This treacherous power entailed the first and most obvious set of special skills in tending and kindling fire – containing it.

The second challenge was presented by ignition. Fire is not only an uncontrolled chemical reaction but, under ordinary conditions, also a nonspontaneous one. Like all nonspontaneous reactions, it requires high-temperature activation energy. To initiate such activation energy on demand (say, by means of friction or percussion) from the materials immediately available in nature is no task for a matchless novice. The discovery of such intricate techniques by people with no prior experience (or need) of handling fire on a regular basis seems highly unreasonable. It is far safer to assume that such a capability was reached only in the course of evolutionary time, after many generations of exposure to the use of fire, and probably not before the first user (*H. erectus*) was replaced by the slightly more innovative descendent (*H. sapiens*). In this respect, the following account and speculation by the archeologist V. Gordon Childe (1951a:47) is as instructive today as it was half a century ago:

> When that discovery was made is uncertain. Savage peoples produce fire by the spark from flint struck against iron pyrites or hematite, by the friction of two pieces of wood, or by the heat generated on compressing air in a tube of bamboo. The first device was being employed in Europe as early as the last Ice Age. Several modifications of the friction method (fire plow, fire drill, and so on) are current among savages in different parts of the modern world, and are mentioned in ancient literatures. Perhaps the variety of methods used for kindling fire indicates that the trick was discovered only relatively late in human history, when our species had already been widely scattered into isolated groups.

In the meantime, before the trick of ignition was discovered, fire had to be borrowed from nature – from lightning or from other natural agency.

To borrow fire from trees set afire by lightning (or from volcanoes), it seems, is easy enough. But then, there is the challenge of keeping it alight round the clock and round the seasons, and occasionally the challenge of carrying it on the road for further propagation. From a purely logistical point of view this kind of undertaking seems even more taxing than achieving ignition. It is this period, in which humans had not yet learned to produce fire at will, that is most interesting from an evolutionary and economic point of view. This period entailed special arrangements in division of labor at different levels of social organization, and left perhaps irreversible imprints on the lifestyle of humankind. I am not sure that our account of prehistoric events always pays due credit to this particular aspect of human history.

Provision of fire in the absence of ignition technology

The following considerations concern the problem of maintenance and propagation of fire in a protohuman society that lacks the technology to start it on demand. The scant body of data available at present about the use of fire by early humans leaves the door open for a number of interpretations that demand careful examination, beginning with the "campfire" scenario.

To entertain an image of our early ancestors as a band of campers clustered around a fire is innocuous enough. To associate this image with the idea of hominid reliance on a central campfire as a primary resource, either for direct provision of energy or for further propagation, is slightly more problematic. It raises a number of intriguing questions both at the technical and at the social levels of organization. First, it should be noted that fires have their optimal size. Small fires are patently unstable and tend to die down of no apparent reason. Large fires are fuel inefficient (and hard to control). Since the temperature at the center of a flame is a function of its size, large fires burn fuel too fast and lose too much heat to the open air above. The family size fire is about the right size. On the other hand, only a fire of much greater magnitude can collectively serve a band of a few dozen hominids clustered around it on a cold night.

Beyond the problem of fuel efficiency, there is a problem of free-ridership. A central fire open to all has all the elements of a public good, as the concept was outlined in Chapter 9. However beneficial to society as a group, individuals willing to undertake the painstaking task of tending a continuously burning central fire, providing it with fuel, protecting it from the elements (and from human errors) – strictly as a voluntary act – are not so easy to come by. Utopian conjectures based on altruism akin to acts of heroism in wartime are noble ideas to contemplate and appropriate in their own setting: catastrophic situations of imminent risk to life (or limb) faced by society at large or by some of its members. But to rely on voluntary action for the purpose of the day-to-day provision of a routine service in the mundane arena of subsistence, is to expect slightly too much of the wrong species in a wrong setting. Short of eusociality, the provision of a public good in the ordinary business of human life must contend with free-ridership. The fact that modern human society still has to contend with this problem indicates that, for better or worse,

humankind was apparently baptized by fire without conversion to eusociality.

A more reasonable alternative, it seems, is the "private fire" scenario:
continuously burning small fires maintained separately by each individual user, typically, a family. The only problem with such a system is how
to accommodate those users who happen occasionally and unluckily to
lose their fires. Short of waiting for the next wild fire, the only practical
solution would have been to borrow it from a neighbor. Initially, neither
a neighbor nor even a stranger should have had any reason to turn down
the desperate borrower. Lending under these conditions is a costless gesture on the part of donors. On the face of it, this arrangement seems to
fit a model of an ideal society where members are driven by proper
incentives to do their best to maintain a reliable fire in their private
quarters and to provide each other mutual, so to speak, "fire insurance"
on demand. On closer examination, however, it soon becomes clear that
the situation is inherently unstable and liable to self destruct.

There is a certain degree of variation in the level of vigilance by which
different individuals handle the material world. Some users of fire are
more vigilant than others. They would put more resources, time, and
effort into the activity, and lose their fire less frequently than others. It
can be surmised that these relatively vigilant firekeepers would have to
lend fire more frequently than borrow it. But so long as borrowing a fire
is free, and maintaining it, expensive, they end up losers in both the economic sense and the evolutionary sense. Others, less vigilant, would
probably have more hours of sleep each night, though, once in a while,
would wake up to a dead cold fire in the morning. They would spend less
time collecting fuel and care less about sheltering their fires from the
elements, thus avoiding a fire-protective distant rock-shelter in favor of
an open site near fresh water and food sources. On rainy or windy days
they would find themselves in a miserable pursuit of donors for a badly
needed "jump-start," but so long as they could get this favor for free, they
would stand a better chance to end up as gainers in a game of subsistence
and survival that rewards free-ridership and penalizes vigilance. Such a
game can not last for long. To put it bluntly, if everyone can borrow fire
on demand, it is no longer in anybody's interest to undertake the
painstaking task of maintaining an ongoing flame. If the players could
not figure out their own interests for themselves in the short run (say,

because they presumably lack rational behavior), natural selection would figure it out for them in the long run. Either way, one ends up with a system comprised exclusively of borrowers, no donors, and no fire. The system, apparently, has never reached such a point of full destruction. New forces were bound to take over and lead to a stable equilibrium of a different kind, as we shall see.

The third alternative, the "incendiary hub" scenario, is probably the most appealing arrangement from a technical point of view. It recognizes the logistical inefficiency associated with jointly consumed large fires as well as the inefficiency associated with a multitude of continuously burning small private fires that must be maintained separately, and continuously, by each individual user. A system of centrally propagated but privately consumed (and fueled) fires is far superior to both. A community of a dozen or so family units using fire for wintertime heating and occasional cooking may cut in half the cost of fuel, far more the cost of labor, and, above all, enjoy a more reliable source of energy, if somehow they could collectively maintain a central fire of optimal size strictly for the purpose of further propagation. Such an incendiary hub could, incidentally, easily serve a population ten or hundred times larger at no extra cost. However, a straightforward solution like this can take no affect without a small group of properly rewarded, thus well-motivated, specialized firekeepers. These kinds of specialists – charging due payment for their services – would undoubtedly have emerged under market exchange. In fact, short of fully fledged eusociality, it is hard to see a social context leading to the formation of such a group of specialized firekeepers without the mechanism of market exchange. True to its special properties (as a contrived commodity), fire was in the ideal position to incite the creation of this very mechanism.

In a world without synthetic ignition devices, any firekeeper can provide fire to others at no extra cost or choose to deny it to those who seek it without payment. In this sense, as we have already seen, fire indeed fits the description of a contrived commodity: a category of goods that is bound to maximize the impetus to trade. This, in addition to the great benefits that a population of early fire users stood to gain from a reliable source of propagation in the presence of an incendiary hub, could well have created the initial condition in the transition to exchange. One can go on and speculate step by step about the dynamics of the process leading from this initial condition to the eventual removal of all inhibitions to trade. What is more important is to recognize that the ultimate selective pressures are bound to lead, one way or another, to the same outcome. My main argument is that the use of fire would have made market exchange highly adaptive, had it not been already in place facilitating the domestication of fire at the outset.

Fire and occupation of caves

Fire is associated, perhaps more than any other sign of early human life, with caves and rock-shelters. For instance, referring to Mousterian/MSA (Middle Stone Age) fossil fireplaces and hearths, Richard Klein (1989:313) points out that they are "a prominent feature in virtually every well-excavated cave where preservation conditions are appropriate." It is also fair to say, as far as I am aware, that almost all the findings commonly considered as reasonably secure evidence for the use of fire by *H. erectus*, early *H. sapiens*, and *H. neanderthalensis* come from caves or rock-shelters. A primary example is Zhoukoudian Cave in north China, a site inhabited by *H. erectus* about 500,000 years ago which contains burnt stones, charred bones, charcoal, and ash beds nearly 6 m (20 feet) deep. Despite some doubts raised by archeologists (Binford and Ho 1985, and more recently by Weiner *et al.*, 1998), Zhoukoudian Cave still provides the most compelling evidence for the possible antiquity of fire in human use. By comparison, open-air sites dating from the last half million years frequently show no signs of fire.[1] The absence of fire traces on open sites is partly explained by the fact that charcoal does not survive well in dry, open air, sedimentary deposits. It can also be argued that open sites, vulnerable to stormy weather as they are, were not ideal places for keeping small fires burning continuously, and hence, the accumulation of charcoal and ash in quantities that could survive for long was less likely to begin with. In any event, the overwhelming presence of fire traces in caves, and the use of the caves themselves as part of the strategy of human subsistence, raise a number of issues that have direct bearing on the subject discussed in the present chapter.

---

[1] To complicate the issue slightly, it should be noted that the oldest (more controversial) evidence for human use of fire comes from open sites (e.g., at Chesowanja, East Africa, dated 1.5 mya).

First, it should be noted that traces of ancient life, and lifestyle, deposited in caves stand a relatively good chance of escaping the ravages of time. Aware of this fact, archeologists concentrate on these sites. Cave representation in the record is, consequently, disproportional to the importance of caves in real historical or evolutionary time. An occasional observer may get the impression that caves were systematically used as standard living quarters for a prolonged period of time or, what is even more doubtful, that entire populations of early humans could rely on cave dwelling as a primary place of residence. Of course, caves have been engaged by humans time and again and, to the present day, they are occasionally used as refuge, as temporary shelter, or for special use purposes such as storage. The issue, however, is whether cave inhabitation could have served human populations as a cardinal solution to the problem of residential shelter at any stage in history. The doubts in this respect come from a number of sources.

From the viewpoint of natural history, caves are a highly random feature of the physical environment. First, they are arbitrarily distributed geographically both relative to sources of food and fresh water, and relative to each other. Second, they are patently idiosyncratic in shape, size, structure, and internal environment (e.g., humidity). As such, they can support the evolution of certain endemic species (that happened by chance to be entrapped in one of them), but have little or no impact on the evolution of free-ranging species. As far as I am aware, no such species uses naturally occurring caves as a key resource in its strategy of subsistence, with the possible exception of certain species of bats. Some animals may use caves for the purpose of hibernation precisely because that is the time they go without food, putting their metabolic and subsistence systems on halt. But even in this limited sense the importance of caves to free-ranging animals is not clear-cut.

Contrary to common preconceptions, large hibernators like black or brown bears do not necessarily rely on naturally occurring caves for wintertime retreat. Occasionally they may use fallen trees or any other object in their range that can provide appropriate shelter. Polar bears dig, for that purpose, their own caves in icy snow. Though remains of the well-known (now extinct) cave bear are often found in cave deposits throughout Eurasia and North America, they most likely used the same opportunistic strategies. It makes little ecological (adaptive) sense for free-ranging animals the size of the cave bear (larger than the grizzly) to limit their geographic distribution only to terrain that can accommodate them in caves. For the same ecological reason, animals that routinely lodge underground typically dig their own burrows or take over those of others in their range. In any event, humans are neither hibernating animals nor burrowing ones. Our closest relatives, the primates, sleep almost invariably on or above the ground, mostly up on trees and in the open air, and are not particularly inclined to suffer from *agoraphobia* (fear of open places). In short, humans are hardly predisposed to colonize caves with any degree of enthusiasm.

Extrapolation from dwelling habits of modern humans casts additional doubts on the legacy of the cave-man. Next to food, shelter is probably the single most important component of human subsistence. In terms of direct cost, modern humans typically spend on housing well above one quarter (though less than half) of total life-time earnings and, as a proportion of personal income, the poor spend on it no less than the rich. Like any item that consumes a large proportion of human resources and effort, we expect – and often find – that residential decisions are highly optimized and finely tuned to the needs of modern dwellers (e.g., Hochman and Ofek, 1977, Ofek and Merrill, 1997). To what extent can we expect different – say, less fine-tuned – patterns of residential behavior on the part of ancient dwellers? To answer this question, some attention must be paid to *opportunity cost* (the cost of a decision assessed by the value of a forgone alternative) as it applies to residential choice. Thus, for instance, if the choice is between taking residence in the business district or moving to the suburbs, then one has to add to the price of a suburban home the full cost of prolonged daily commutation.

Many factors affect the choice of residential location by modern people, chief among them is the distance to the place of employment. Indeed, the most expensive item in transportation for most people is the daily round trip to work, if not in terms of out-of-pocket cost, certainly in terms of travel time. A typical commuter with a job in Manhattan and a suburban residence, say, 50 miles away on Long Island, will annually cover a commuting distance equivalent to a trip around the world, spending what amounts to 80 working days on the road (just enough to win a Jules Verne bet) – not counting traffic jams. Patterns of residential location in modern societies reflect the relentless attempts by households to minimize the home to job distance (subject to the constraints of income, real-estate prices, and several additional factors such as the availability of quality public education, in the case of families with school-age children). Akin to models of optimal foraging behavior in animals, the problem of commuting and optimal residential choice is not the easiest textbook exercise in economic location theory. However, most people seem to solve it intuitively to the satisfaction, and sometimes surprise, of experts. This may suggest ancient origins.

Indeed, as much as we sympathize with the predicaments of modern commuters, prehistoric transportation posed a slightly more difficult challenge to our ancestors. If nothing else they had to travel by foot and carry their take of the day on their shoulders. A mile was a measure of energy rather than distance. Any small error in residential location, a mile or two off target, would mean the equivalent of hundreds of miles of lost energy per year. Energy expended on travel and transport in long-range hunting-gathering can easily exceed the energy intake from the kill or the find on any particular day. When the energy balance is upset, in this sense, on an annual or seasonal basis the results can be devastating.

Beyond motorized transportation (or draft animals), a modern observer should be aware that running water and electric power are more than a matter of convenience. The lack of both entailed two additional daily trips in ancient dwelling: one to a source of fresh water and, if fire was in use, another to a source of fuel. In short, our ancestors faced a problem in optimal location subject to three spatial constraints (compared with one or two in the case of the modern household, depending on the number of earners). Location theory might guarantee a well-defined solution to this problem: the existence of an optimal point in the landscape at which one is well advised to locate. Nobody can guarantee, however, the existence of a cave at this particular site, nor even within walking distance from it. In the unlikely event that a cave would occur within a walking radius, if only 2–3 miles away, it may still be more sensible to put the one-shot extra effort of constructing an open site shelter at the optimal spot. It would save about 500 walking miles per season (or 4 months) for a nomadic hunter-gatherer, or (assuming a 20 year span of working life) as much as 30,000 miles for a sedentary one – more than enough for a walk around the globe. In their version of a real-estate market, where rents were paid by feet, a cave was a dwelling the average hunter-gatherer could hardly afford.

There is no need to throw out the baby with the bath water. Ruling out commonplace cave dwellings does not preclude practical use of caves in other ways. There are undoubtedly examples of caves in occasional or temporary use as a home-base for certain individuals or groups even in recent times. However, this sort of cave occupation has typically had little economic bearing at the level of populations. There is still a possibility that at certain stages in the course of human evolution special-purpose cave use (other than dwelling) was important to the survival or welfare of human (or proto-human) populations at large. Granted a stage in human prehistory that fits this description, what can we say about the people who actually occupied these caves? Suboptimally located with respect to food sources, at least some occupants of these caves would have to rely for various parts of their subsistence on transfers from the rest of the population, arguably, in return for some service. At least some would have to be engaged in an activity that deviated from the ordinary (the "ordinary" being hunting and gathering). This activity should have been performed at clear comparative advantage due to some feature of caves unshared by open sites.

The only advantage over an open site shelter that a cave can offer (except for saving in construction cost which, as we saw, could not have been a very compelling consideration) is a relatively secure place for kindling and maintaining a continuously burning fire. A cave obviously provides better protection from the elements for the fire itself and, in addition, an ideal (relatively) dry storage place for fuel and tinder. The only caveat is the fact that the fire handler cannot subsist on fire alone. What we face here is a situation that suggests a degree of specialization along with exchange: the existence of specialized firekeepers. The basic conditions and system of incentives for the emergence of such specialized agents in a society that makes use of fire, but lacks appropriate ignition technology, was discussed at length throughout earlier parts of this chapter, albeit, in a way of theoretical expectations. It is the prevalence of ancient fire traces in excavated caves, sometimes in unexplained abundance, that should lend some empirical support to these theoretical expectations.

The main conclusion from the discussion so far leads to a number of useful predictions about the possible use of caves and the subsequent development of ignition technologies. It suggests a long cycle of rise and fall in cave use. When fire was first adapted for human use, assuming market exchange was not yet available, the use of caves for kindling fire must have been fairly limited. In fact, as I have already mentioned, the earliest (and more controversial) indications of possible human use of fire come from open sites (e.g.. at Chesowanja, dated 1.5 mya), rather than from caves. With the emergence of exchange along with the conditions for specialization and incentives for division of labor, the use of caves precisely for the purpose of an "incendiary hub" probably intensified and eventually reached its peak just before the introduction of ignition technologies.

It is reasonable to assume that at this point the cave became a focal point of social activity at least in parts of the world where caves occur with reasonable abundance, typically, limestone and karst regions. Through agglomeration (the tendency of enterprises or activities to locate near one another) an economic activity that takes place at a particular location where it enjoys a clear comparative advantage is generally bound to attract to the same site other activities – cultural activities included. Cave art is probably a primary example of such a collateral activity, but not necessarily an exclusive one. Stringer and Gamble (1993:167) mention, for instance, the possibility that caves in conjunction with fire were used throughout the glacial northern latitudes as a convenient place to raise the temperature and defrost frozen game or carcasses of animals that died by natural attrition. Apparently, while the meat was thawing, the artists in residence could safely draw scenes of wildlife in action on the walls right next to their still models.

With the introduction of ignition technologies, the decline of the cave as a specialised fire depot was inevitable. I do not know if the evidence collected so far would allow one to draw temporal inferences about the intensity of cave use with any degree of certainty, but, on purely theoretical grounds, one might anticipate seeing a fairly well-defined historical period during which caves were used with higher frequency and intensity. If this were the case, then the period in question would roughly match an interval in time starting with the earliest use of fire-augmented exchange (or with exchange-augmented fire use) and continuing through to the introduction of ignition technology and its first widespread use.

11   The upper paleolithic and other creative explosions

Modern humans stayed anatomically unchanged at least for the past 80,000 years. On the evolutionary level of organization, anatomically fixed things are expected to stay (nearly) fixed in behavior. Our (anatomically) modern ancestors lived up to this rule for the first half, or slightly more, of their tenure on earth. All hell broke loose in the second. The extraordinary changes in the archeological record starting around 40,000 to 30,000 years ago, and carrying through the height of the last ice age to the onset of the Holocene (some 10,000 years ago), suggest remarkable refinements in behavioral structures unexpected of a morphologically fixed organism. Changes in the record further suggest a remarkable increase in regional and temporal diversity of material structures that up to that point varied little through time and space. The Middle to Upper Paleolithic transition, or the *creative explosion* as this episode has sometimes been labeled (e.g., Pfeiffer, 1982), is most vividly evident in wall paintings preserved in caves, in portable art, personal ornamentation, and in elaborate burials. More subtle are the sudden refinements in tools, and the rapid expansion into new geographic areas, indeed, into two new continents (Australia and the Americas). Underlying all of this is an authentic economic expansion reminiscent of various mercantile and industrial revolutions in recorded history.

The key question, from an evolutionary viewpoint, is how could such remarkable changes take place in functional behavior without apparent change in morphology. One possible explanation ascribes this turn of events to some neurological change that led to an evolution in behavior without an apparent change in anatomical form (e.g., Klein 1992). Alternatively, it has been argued that on this occasion "culturally organized behavior ... revolutionized our evolution in a way that may have been quite independent of genetic change" (Binford 1992). Choosing between such ultimate explanations is beyond the reach of current observations and will probably remain a tantalizing unsettled issue for some time. In the meantime, I would like to side-step the controversy without completely evading the issue by trying to subject the factual details, over which there is relatively little disagreement, to the scrutiny of economic reasoning.

The Upper Paleolithic toolkit

Scatters of ancient stone artifacts are best viewed as the tip of an iceberg. Hidden and no longer recoverable is a larger technology including tools and implements made of wood and other organic perishable materials (hide, fiber, wool, ligament, etc.). Further hidden and far more subtle (even to observers in true time) are highly interwoven structures of division of labor and specialization in society. If we look with some care at various items in the Upper Paleolithic toolkit, noting their refinement and deployment within and across populations, they soon seem to suggest improvement in market structures, division of labor and, quite possibly, a major underlying innovation in trade. Four observations regarding Upper Paleolithic tools seem, in particular, to reinforce the notion of intensification in division of labor and specialization:

(1) New tools geared for specific tasks replaced general-purpose tools.
(2) The new tools were more expensive to manufacture.
(3) The new tools were more difficult to master.
(4) The new tools show a remarkable increase in regional variations.

The replacement of general-purpose tools with tools geared for specific tasks is evident, for instance, in the appearance of specific hunting weapons designed for specific types of game. The extra cost in manufacturing is evident in a number of ways. Unlike the earlier (Mousterian) stone tools technology which was dominated by *flakes*, the Upper Paleolithic technology is marked by well-made *blades* (flakes at least twice as long as wide) that require special effort and skill to produce routinely. The Upper Paleolithic period is also marked by an increase in the use of hafted or otherwise composite tools and implements (combining multiple components made up of different materials). Composite tools exact extra assembly costs of fitting and binding together naturally occurring – and thus inherently unstandardized – components. Moreover, the period is marked by widespread use of tools made of materials largely ignored by earlier people: bone, ivory, and antler. Though lighter and often more durable than similar tools made of stone or wood, they also required a larger expenditure of time and effort in processing.[1] But the most direct testimony to the Upper Paleolithic people's willingness to make sizable investments in tools comes from a class of artifacts they left in the record in relative abundance: tools for the creation of other tools.

The difficulty of mastering a tool is measured by effort in training. The best judge in this matter is the apprentice. Every apprentice in a carpenter shop, for instance, is probably well aware that it takes longer to master a saw than a hammer. A hammer is essentially a protracted hardened fist, slightly augmented in weight, which serves as a simple extension of the human hand. In this sense, it greatly differs from a saw. The primary function performed by a saw is more nearly embodied in the design of the tool than in the design of the hand that drives it. The same is true of a needle: a minor utensil by modern housekeeping standards, it was an innovation of considerable importance to northern people of the last ice age (see Figure 11.1). The saw (and blades in general) as well as the needle belong to a class of truly synthetic tools in the sense that they have no close analogue in any part of the human body.[2] They present new tricks to the neural system that coordinates manual dexterity – tricks for which our brain is not hardwired. This generic characteristic of tools, perhaps more than any other, is the hallmark of the items that made their first appearance in Upper Paleolithic toolkit. Harpoons, bows and fish nets (perhaps), traps for snaring animals, sledges, grinding stones, seagoing rafts or boats (presumably used by first arrivals to Australia), bone needles, and indeed, the very microburin drills that pierced eyes in these needles – are some of the examples in this class. Tools that require new brain circuits, and all too often new strategies, take longer practice to master.

Finally, the extensive spatial and temporal variability observed in tools

---

[1] Climate was also a contributing factor in this transition to new materials. With the approaching glacial maximum (20,000–16,000 years ago) certain types of wood became more and more scarce due to widespread deforestation. In addition, bone and antler tools could stand the cold better than stone tools (which tend to become brittle in low temperatures).

[2] The particular origin of the saw itself (i.e., the main idea in its design) can be traced most directly to the microlithic tradition, especially to the reaping knife and sickle blades, which in Europe are more typical of the *Mesolithic* period (about 10,000 to 6,000 years ago, but earlier in the Middle and Near East). However, blades in general are the undisputed domain of the *Upper Paleolithic*.

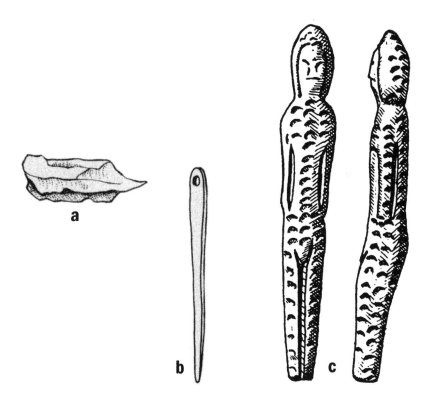

Figure 11.1 **An elementary tool for the creation of tools and some of its effects** A late Paleolithic microburin drill for piercing holes in skin, wood, bone, or antler (a) made possible the production of eyed bone needles used to sew skins together (b). The needle facilitated well-tailored airtight protective clothing essential for human colonization of the frozen north during the fiercest stage of the last glaciation (adapted from Lambert, 1987). The early existence of such sophisticated garments is evident from the fur-clad individual in the carved figurine (c). Made from mammoth ivory some 20,000 years ago, it is also an indication of the degree of sophistication in prehistoric artistic imagery (found at Buret, north of Irkutsk in Siberia, as depicted in Derev'anko, 1998, Fig. 114). More direct evidence for the existence of well-tailored clothing comes from Upper Paleolithic burials (especially at Sunghir, Russia – a site dating 23,000 years ago – where strings of beads and other clues suggest the details of hats, shirts, trousers, and moccasins made of fur or leather).

and other artifacts of the Upper Paleolithic record stand in sharp contrast to the uniformity of earlier periods. For instance, for more than a
million years and regardless of continent and latitude, one can hardly
see any change in the basic design of the hand axe (perhaps the most
popular tool ever and certainly one that stayed longest in human use).
Indeed, over long time spans and across vastly separating distances, the
predecessors of the Upper Paleolithic people seem to have formed their
stone tools from the same common template. Everything changed, however, 40,000–35,000 years ago with the appearance of the remarkably
diverse toolkits of the Upper Paleolithic, containing well distinguished
tool types and subtypes not only across a wider range of specialized activities, but also across sites and regions.

Taken together and each separately, the first three of the four observations listed above clearly suggest tools designed by and for specialists,
or at least for semi-specialists. Compared with the previous record, they
thus suggest a sudden increase in the level of occupational specialization
and division of labor. The last observation suggests that the new phenomenon was not confined at the level of local communities but extended in space across them. By implication, we can infer a corresponding
increase in the volume and scope of exchange not only within but also
across the Upper Paleolithic societies.

Long-distance trade

There is little doubt that long-distance trade was part of the Upper
Paleolithic way of life, as evident from a multitude of objects found far
from their original sources. Seashells from the Mediterranean appear at
Upper Paleolithic sites several hundred kilometers north in central
Europe; fossilized amber from the Black Sea is found in central Russia (up
to 700 km away); and trade in seal skins along the Atlantic coast was
inferred from seal skulls that occur with no other bones in Spanish caves
far south of the range of this particular species (presumably the skins
were traded with the heads attached; Linton, 1956). But the most compelling examples are associated with the production of tools from certain
types of stones that were routinely transported 100–200 km (and up to 400
km for more distinctive high-quality flint) from Upper Paleolithic quarry
sites in north central and eastern Europe (Klein 1969, Schild 1984).

Long-distance trade leaves clear marks not only at its final destination but, occasionally, also at origin. For instance, indications that fur-bearing animals (especially wolves and arctic foxes) were caught for their skins is evident from skeletons (lacking paws) found in at least half a dozen Upper Paleolithic sites in Russia and Ukraine. (Trappers often remove the skins with the paws attached, and then discard the paw-amputated skinned carcasses.) From the extraordinary abundance of these remains it is reasonable to infer that the furs were used for a purpose beyond strictly domestic consumption, suggesting the obvious possibility of longer distance fur trade.

All trade begins or ends with some local transactions, long-distance trade included. But when long-distance trade is carried out with some regularity, it is safe to assume that trade in purely local goods and services (e.g., perishable foodstuff, housing, and, especially, labor services) has already reached a considerable level of intensity.

Economic and geographic expansions

Further indications of a considerable increase in the intensity of exchange – both local and long distance – are suggested by two additional observations, one associated with an overall economic expansion and the other with a nearly global geographic expansion of the Upper Paleolithic population.

Richer and more extensive excavation sites than earlier (i.e., Mousterian) sites in the same general areas suggesting far better built dwellings and an increase in the density of populations, skeletal remains implying a far reduced incidence of serious injury, disease (e.g. dental hypoplasia), environmental trauma, juvenile mortality, and a general increase in longevity – all point to rising standards of living as part of a substantial Middle to Upper Paleolithic economic expansion. Upper Paleolithic people apparently used local resources more efficiently than their predecessors – or their Neanderthal neighbors – if the latter still existed as a separate entity at the time (Klein, 1989). Such a sudden increase in the "wealth" of populations suggests a corresponding improvement in the allocation of resources in society, most likely, in my opinion, through the mechanisms of division of labor, exchange, and investment in the most consequential resource of all: *Human Capital* (Box 11.1).

Box 11.1 The Upper Paleolithic stock of human capital
Human capital is the idea of human beings investing in themselves through accumulation or optimization of knowledge, skills, good health (hence longevity), geographic location, etc. (developed primarily by Gary S. Becker, 1993, and Jacob Mincer 1974). Like all investment, the accumulation of human capital is a balancing act between expenditure of resources in the present and returns in the future. Returns to human capital come mainly in the form of increased productivity and the capacity to utilize more efficiently resources in consumption (both multiplied by the effective life span and properly discounted in time). Next only to schooling, a major contributing factor to the stock of human capital is the level of training and occupational specialization. The limiting factor is (untimely) mortality.

On both counts, the stock of human capital per capita seems to have risen substantially in the course of the Middle to Upper Paleolithic transition. Training and occupational specialization appear to have intensified for reasons already outlined in the text. Concerning rates of mortality, archeologists point out that the Upper Paleolithic life expectancy exceeded that of the Neanderthals, perhaps by as much as 20% (Klein, 1989; Soffer, 1994). It should be noted that Neanderthal males rarely survived to their late forties and, for reasons associated with childbearing, females rarely survived to their late thirties; that is, roughly the ages at which human capital tends to reach its peak. It follows that the stock of human capital actually had to increase by a factor more than 20%.

The Upper Paleolithic (late-Pleistocene) people greatly extended the geographic distribution of humankind to include easternmost Europe, northern Asia (Siberia), Japan, Australia, and the Americas. Indeed, by the end of the period, humans were present on all continents except Antarctica (Figure 11.2). What could possibly spur a sudden wave of migration of such remarkable magnitude? The fact that the continental shelves were exposed during the Upper Stone Age (due to falling ocean levels in the course of glaciation) is undoubtedly an important contributing factor, but can hardly provide a full explanation. The conti-

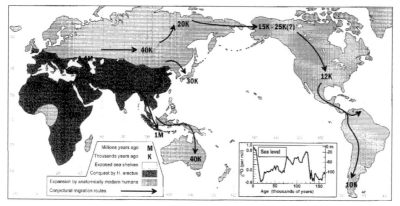

Figure 11.2 **The worldwide spread of human populations** Dated fossils
and artifacts that occur in various parts of the world suggest a general
pattern of human dispersal out of Africa, commencing with the early
migration of *Homo erectus* (about 1.4 million years ago). Considerable
acceleration in the process can be detected (starting about 40,000
years ago) with the approach of the height of the last glaciation and
the geographic expansions of modern humans to higher latitudes
(Siberia) and to new continents (Australia and the Americas) and
islands (such as New Guinea and Japan). As they came to occupy new
destinations, anatomically modern humans somehow replaced their
human antecedents at all previous destinations to form one and the
same species throughout the world. Whether this evolutionary
transition from *Homo erectus* to *Homo sapiens* was primarily
accomplished through (unidirectional) repeated migrations out of
Africa and physical replacement of local populations, or primarily
through local evolution supported by extensive (multidirectional) gene
flow is, at present, a largely unsettled issue (and a point of lively
contention between the Out of Africa and the Multiregional schools of
thought) (insert: after Raynaud *et al.*, 1993).

nental shelves stood exposed on many previous occasions since the first
dispersal out of Africa (estimated at 1.4 million years). This includes a
dozen or so episodes of maximum glaciation when the exposure reached
levels equivalent to the Upper Paleolithic peak. Ample opportunity for
human migration to nearby islands and distant continents was thus
available on many prior occasions. Moreover, with eastbound migration
flowing from Upper Paleolithic central Europe toward Asia and northbound migration moving on both continents toward the arctics

(seemingly heading in the wrong direction in the midst of an ice age), the major thrust was largely inland rather than overseas. The role of exchange suggests itself.

Depending on intensity and scope, trade can be an incentive and a powerful vehicle for migration, range expansion, and long-term occupation of harsh environments. To understand why, one must first recognize that there is a fundamental difference in the manner in which human populations move geographically under exchange, and in its absence.

Without exchange, a population (as such) could never make a sustainable move into a new geographic area unless the full configuration of resources necessary for its long-term survival is locally available. The range and mix of resources needed to sustain human life in the long run exceed by far most, if not all, other mammals for three fundamental reasons: (1) reduced gut volume, as we saw in Chapter 5, imposes unusually stringent diversification requirement on human diets; (2) dependence on raw materials from the physical environment (for tools, clothing, and shelter); and, (3) dependence on fuel, especially in the cooler areas of the world. Such a configuration of resources will geographically occur only by coincidence even in tropical or subtropical regions, let alone the harsh environment typical of the Eurasian Upper Paleolithic. The margins for error or failure in geographic mobility are far too slim for sudden bursts of mass migration to take place under these conditions. Nor would there be, in the absence of trade, much incentive for long-distance exploration on the part of individuals. A new discovery of a treasured resource in a remote area would mean little to the lucky traveler who came across it. Capitalizing on the find would depend on the ability to bring together the right number of consumers for it and the local availability (again by coincidence) of all the resources necessary to support them there, since import and export are out of the question in the absence of exchange. Geographic mobility under these conditions will probably operate not by migration as we know it, but by radiation: a painstakingly slow process typical of plants and even animals in which a population (or a forest) moves incrementally by minor modifications in range, generation to generation, but each individual in each generation stays virtually put. Indeed, in the absence of exchange, humans because of their eclectic vital needs in consumption are doomed to move (as populations) by the laws of motion of a forest, and without the advantage of aerial dispersal of seeds.

Box 11.2 Long-range trade by radiation

Imagine yourself a member of a society that lacks mechanized transportation or beasts of burden. The following (numerically and otherwise highly hypothetical) example illustrates an opportunity in the business of long-distance trade that you may wish to pursue. If you possess a fox fur, and a fox fur is traded for only 20 dried fish 4 miles up the road to the north but for as much as 21 a piece the same distance to the south, and if you are willing to walk 16 miles each day, then you can have every night a fish for dinner and still remain in possession of a fur (though not of the same fox). If this is the case, then you can consider yourself a well-qualified long-range trader with little need for long-distance travel or any particular skill of social interaction with foreigners – and you will never miss a night at home. You are, no doubt, highly conscious of fur and fish prices in local markets, but reduced to guessing about where the stuff is coming from and where it is headed. Not that you care about such things, but you may be surprised to learn that some trappers 80 miles up in the mountains to the north are willing to catch foxes (perhaps the only resource in their range) and skin them on demand for as little as 10 fish a head. It is quite possible that without these transactions, which supplement their food supply, the trappers would fail to survive in their range and probably would have never settled there to begin with. (So much for the key role exchange plays in the dispersal of people over the face of the earth.) You will be equally surprised to learn about the existence of a seashore another 80 miles down to the south where fishermen are willing to acquire the same skins for as much as 30 fish a piece. The existence of some 20 traders stringed and roughly equally spaced along the same road and all, just like yourself, driven to action by spatial price differences should perhaps come to you as no surprise – and you can probably guess what each will be having for dinner.

The topography and numbers are strictly fictional, the characters are out of context, but the fundamental principle just outlined in this embryonic model is fairly applicable to any system of spatial distribution that ever operated under the laws of supply and demand.

On the other hand, under exchange, human migration is no longer conditioned by regional resources in any particular configuration of diversity. The availability of a single resource or commodity can spur voluntary migration to the most desolate of locations, and so long as there is sufficient demand for it somewhere in the system, however remote, the new inhabitants are going to be provisioned by long-range trade – the ordinary traffic of imports and exports. It should be noted that the system of distribution will establish and regulate itself spontaneously in response to little more than differences in local prices (see Box 11.2). Thus, trade clearly promotes migration and facilitates the dispersal of human populations into harsh new environments that, otherwise, would have never been settled.

There are a number of implications of special bearing upon the Upper Paleolithic setting. If trade played a role of any importance in the geographic distribution and dispersal of these ancient human populations, we would expect to see an increase in regional specialization. Regionalism in this sense is, no doubt, one of the hallmarks of the Upper Paleolithic; so much so, that there are even indications of the exploitation of regional specific resources intended primarily for export. Quarries for export of flint is an obvious example. I already mentioned the possibility of trade in seal skins along the Atlantic coast as well as trade in skins of fur-bearing animals (especially wolves and arctic foxes) further to the east across Upper Paleolithic sites in Russia and Ukraine. It should come perhaps as no surprise that the artifacts found in one of these sites (Mezhirich, near the Dnieper river in the Ukraine, dating to a period over 18,000 years ago) include jewelry made of shells from the Black Sea (300 km away), amber from the Baltic (900 km), and tools made of imported stone. This particular site is also known for its extraordinarily large accumulations of mammoth bones collected deliberately or used opportunistically, for building houses, presumably. For reasons discussed in chapter 8 (in connection with the Oldowan assemblages of unprocessed stones), enigmatic accumulations of distinct commodities or materials are to be expected occasionally at archeological sites, assuming the ancient inhabitants of these sites were engaged in trade. Trapping and the fur trade, it seems, have always been associated with human expansion into the northern frontiers. Until quite recently it actually led the penetration and colonization of Siberia and North

America. In fact, the very first cargo shipped out of Plymouth, Massachusetts (which departed on September 10, 1623, bound for England), consisted of lumber and furs.

Monetarization of exchange in relation to symbolic behavior

A general view shared by many archeologists is that (perhaps with the exception of the *Australopithecus* to *Homo* transition 2.5 to 2 million years ago) the transition from the Middle to the Upper Stone Age is the "most dramatic behavioral shift that archaeologists will ever detect" (Klein, 1992). Others see a parallel between this transition and the transition to agriculture, suggesting that it will be methodologically useful to adopt a similar approach to both (Bar-Yosef, 1994). I would go a step further and, for whatever it is worth, suggest that the analogy might be extended to the Industrial Revolution, for which we have a well documented written record and in which we are, in a way, still living. At any rate, from the standpoint of economic history, it is not unreasonable to view the Upper Paleolithic transition as one of the first in a series of fairly successful human attempts to escape (as populations) from poverty to riches through the institution of trade and the agency of the division of labor.

One lesson we learn from the Industrial Revolution, and from similar expansions in recorded history, is that they rarely take place unless the economic sphere of operation reaches a certain degree of autonomy: a set of conditions that enable individuals, as such, to make decisions at their own risk, take independent action and responsibility for the outcome, as well as earn the returns or suffer the consequences – all under reasonable freedom from arbitrary confiscation of property rights. Some economists, perhaps most (myself included), trace the origin of the Industrial Revolution to the creation of such a free economic sphere (e.g., Rosenberg and Birdzell, 1986). The Industrial Revolution produced, and in a more fundamental way, was a product of a major innovation (or renovation) in the institutions of trade: the modern corporation. Under proper protection and regulation of the law, the corporation is an institution that can dramatically improve the mechanisms of division of labor through shared ownership and recombination of capital. It was not, however, the first major innovation in trade since the onset of exchange. It was preceded by a more consequential innovation which possibly belongs, I will now argue, to the free economic sphere of the Upper Paleolithic transition: the introduction of money in some form (pre-numismatic, to be sure) as a medium of exchange.

One indication for widespread use of money during the Upper Paleolithic period is the sheer volume and intensity of exchange that can be directly inferred from the evidence of long-distance trade (i.e., the visible tip of the iceberg), or indirectly so, from much of the other observations just outlined. The most compelling argument is associated, however, with a body of evidence that was mentioned, but so far has received little attention in the present discussion: the appearance of wall paintings, portable art, personal ornamentation, elaborate burials – that is, the sudden increase in the level of symbolic behavior. The use of money is certainly a form of symbolic behavior, the only form of symbolic behavior for which we have a straightforward adaptive explanation and, as such, quite possibly the origin of all symbolic behavior.

By its very nature, money is a fiction (in a way, so too is the corporation) and dealing with it is the ultimate ritual. Money, generally, has no value in and of itself (nor does the corporation).[3] Its value is unrelated to its material existence or its merits in any use but exchange itself. For instance, money exists today mainly as "digital cash" (i.e., electronic impulses stored on magnetic tapes or computer chips) that make little sense and have no use whatsoever to their owners. Almost everybody is willing to accept these senseless electronic impulses in discharge of claims merely because they are acceptable by almost everybody else in the same manner, and precisely for the same reason. By virtue of this mutually accepted fiction, money becomes as real as the most tangible commodity it can buy. It is this willingness of people to accept things that they do not plan to use except for passing them on to others that greatly increased the efficiency in exchange. Money differs from all other commodities in one particular sense. The appraisal of an ordinary commodity is its value in self-use – which is fairly easy to figure out. The

---

[3] A seeming exception is *commodity money* (gold, shells, dried fish, fox furs, etc.) which has intrinsic value. However, when such commodities start to function as a fully fledged form of money (a medium of exchange, store of value, unit of account, etc.), their intrinsic value may fall arbitrarily short of their market value and is no longer essential to the function they perform as money. As to the corporation, all its equity, to the last penny, belongs to share-holders.

appraisal of money, on the other hand, is its value to others – for the most part, anonymous recipients – which requires some abstraction and the capacity for symbolic thinking.

Paradoxically, the use of money which is driven by the most utilitarian of motives, shares a number of common characteristics with different forms of ritual behavior, as well as with representative and fictional forms of art. We can expect therefore a relatively high level of symbolic behavior on the part of members of a society where money (numanistic or otherwise) is in widespread use, and vice versa. By implication, it can be argued that the Neanderthals had very little use for money. Though they were probably engaged in some forms of barter (which I take the liberty of inferring from their use of fire), they show very little in the way of art or symbolic behavior compared with their Cro-Magnon neighbors. Either they were lacking the mental capacity of abstraction and symbolic thinking necessary for handling money, or they were simply confined to pockets in the environment absent the necessary conditions (which I will shortly discuss) for the incitement and development of this practice. In any event, the lack of this practice could have put them at a substantial comparative disadvantage and, as such, may have played a role in their final demise.

If we accept, if only for the sake of argument, the possibility that the use of money by modern humans was introduced, or greatly intensified, sometime during the Middle to Upper Paleolithic transition, we may have part of an answer to the (proximate) question of "How" the transition took place, though thus far no clue as to the (ultimate) question of "Why" did it take place to begin with, and "Why" then and there and not before and elsewhere. In view of the apparent morphological stasis, these "Why" questions are the most important of all. One obvious possibility is the occurrence (then and not before) of an external independent triggering event in the physical environment. The Upper Paleolithic conspicuously coincided with such an event: the approach and onset of maximum glaciation (of 20,000–16,000 years ago), the first of its kind to be experienced by anatomically modern humans. What I will argue in the remaining part of this section is that the larger trend associated with the Upper Paleolithic transition can be traced to this environmental event.

It is tempting to view the ice age as a great hardship, and the peak of the ice age (maximum glaciation) as maximum hardship. There is undoubtedly a great deal of justification to this view at the level of the individual. An individual, as such, stands a greater chance of perishing under conditions of glaciation than otherwise. And if one does not perish, there is not much to enjoy in the way of amenities in an environment where the wind chill, rather than fire, is the main cause of continental deforestation. At the level of organization of populations or species, however, there is no such thing as absolute bad weather. As I have already mentioned, and now is the time to elaborate, the worst weather from the point of view of an individual might be a time of great prosperity to the population or the species, if the competition is affected more adversely. To humankind, the Upper Paleolithic glaciation was a time of great prosperity, precisely for that reason. At issue is expansion at the expense of competing predators.

Almost everything one needs to know about primate safeguards against predation on open grounds can probably be learned, as we saw (in Chapter 7), from the ecological plight of the baboons. Though the number of baboons that fall prey to predators is remarkably small, predation remains a major limiting factor in their ecological adaptation to grassland. Predation rates are low precisely because the expenditure on anti-predator safeguards consumes so much resources and energy that the entire social organization of the baboons, their reproductive behavior and ranging, all hinge on this particular need. One of the most effective, and most expensive, structures that the baboons developed in order to cope with predators is the close group formation they always maintain on open grounds. Protection against predators is thus achieved at the cost of individual freedom in ranging.

During the last ice age, the large carnivores (lions and other big cats, as well as hyenas), which prey on humans and their game, were present not only in Africa and parts of Asia, but also throughout much of Europe. Before the invention of the bow, if not firearms, it is reasonable to assume that humans were much in the same predicament as the baboons relative to predators in their range, and probably had to deploy similar antipredator measures. The only safeguard against the large predators on open grounds and away from shelter, especially in the approach to sources of fresh water, was movement in formations consisting of several reasonably armed individuals. The constant need to move in group formations was quite likely a compelling constraint in the strategy of human subsistence in areas shared with large carnivores. The size and configuration of such formations, as well as their group dynamics, were probably determined more by safety considerations than by the task at hand. The most obvious costs, from the viewpoint of the hunter-gatherer, were forgone opportunities in individual ranging: all benefits they stood to gain from activities best performed independently by individuals (e.g., hunting fast-moving small mammals such as hares or hedgehogs, trapping, fowling, fishing, and the exploitation of plant food that occur in widely dispersed small patches). In the absence of free ranging on the part of individuals a society, for its part, can hardly achieve a reasonable degree of optimal distribution of human resources in geographic space. In a more fundamental sense, the need for collective safeguards against predation and other life-threatening contingencies (warfare and violent crime) comes always at the expense of certain individual freedoms. Among them is the freedom to take independent action, freedom of association (and dissociation) which is important to the formation of task-oriented voluntary teams, and the freedom of one-on-one interaction important for the conduct of exchange in the relative safety of the marketplace. Such freedoms are hardly promoted in a society that must constantly conduct its business in battle formations.

Glaciation changed this reality by conferring two major advantages on humans. Deforestation increased the range of grazing game, and falling temperatures reduced the range (or at least the presence) of large carnivores that prey on it, and on humans. At issue are some specialized (i.e., truly carnivorous) predators larger than the wolf: certain hyenas and, especially, big cats like lions. Despite their formidable size, bears generally posed a far lesser threat.[4] Both humans and big cats are essentially

---

[4] Though classified as carnivores, bears appear to be unaware of their place in taxonomy. They are all omnivores – the polar bear and the (now extinct) cave bear, included. Lacking the sheering teeth common to carnivores, bears subsist for the most part on plant food and insects, occasionally on fish, and only rarely on mammalian prey (in this respect the polar bear is the exception). Bears in natural settings tend to avoid or ignore human presence for a good reason. Summertime, when plant food is in abundance, stocking or ambushing unpredictable prey must be a waste of time and energy for a vegetarian omnivore. Wintertime it hibernates or stays inactive posing few problems to humans. In more recent historical settings, bears probably served humans more as game than as a threat of predation (e.g., Iregren, 1988).

derivatives of tropical and subtropical woodlands and savannah that found themselves competing for the bounty of newly formed vast grasslands (ungulate game and prey) in the subarctic regions of the world. Without special adaptations, both would have a hard time withstanding a wind-chill factor cold enough to cause deforestation (which made room for grassland to begin with). Granting evolutionary time sufficient for speciation or subspeciation, northern wildcats, like the (now extinct) European lion, were apparently quite capable of growing long dense furs and developing the necessary anatomical and physiological adaptations. Such adaptations sustain to this very day the lynx (*Lynx linx*) in the high latitudes of northern Europe and Asia, as well as the snow leopard (*Panthera uncia*), the mountain wildcat of the Himalayas.

However, fire, protective clothing, and construction of dwellings, gave an edge to humans. With the intensification of the ice age, approaching glaciers from the north and the expansion of glaciers further to the south along much of the Alpine-Himalayan orogenic range of mountains stretching from North Africa through Europe and Asia toward Indonesia (including the Pyrenees, Alps, Caucasus, Himalayas and almost everything in between and beyond) established a northern ice-free corridor almost impenetrable to wildlife from the south during the height of glaciation (see Figure 11.3). Much of the Middle to Upper Paleolithic transition took place in this ice-free corridor stretching between the Scandinavian ice sheet and the Alps in Europe and between the shores of the Arctic sea and the Himalayas in Asia (Figure 11.4).

We now know that the climate of the ice age was far more volatile than any expert would have believed only a decade ago (more on this in Chapter 12). Repeated episodes of local extinction caused by devastating and unpredictable sharp spikes of Arctic force weather, typical of this climate, probably afflicted many species. Thanks to the synthetic adaptations mentioned above, human populations evidently withstood the extreme harsh climate without becoming extinct. Large carnivores like the European hyenas and lions were more vulnerable. Undergoing an episode of local extinction such carnivores, like other northern species of tropical or subtropical descent, could have been replenished by their southern stock only with great difficulty and long lags in time. Depending on the distance to the nearest mountain pass that somehow stayed open under glaciation, such lags in time could last for decades if

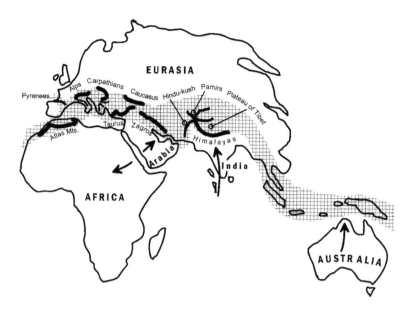

Figure 11.3 **The Alpine-Himalayan Orogenic belt of mountains** The continuous band of mountains stretching from North Africa and the Iberian peninsula, across Europe and Asia Minor, to India and Indonesia, is a product of compressive deformation ascribable to plate tectonics. As continents crumpled against each other adjacent rock bodies and seafloor material with nowhere else to go produced immense compressive forces resulting in great vertical uplift (up to 7 km) and subduction (up to 30 km below the normal depth of the earth's crust). In a protracted process starting some 66 million years ago parts of north Africa swung northward towards Europe to produce the Alpine mountain systems, Arabia and Iran collided with Asia to produce the Taurus, Zagros, and other mountains of Asia Minor, and India collided with Asia giving birth to the Tibetan plateau and the Himalayas. This process of mountain building culminated roughly between 30 to 10 million years ago, but in many parts the process continues to this day. For reasons outlined in the text, the Alpine-Himalayan combined range of mountains is one of the most consequential geomorphic structures in the history of anatomically modern humans, especially in the last two stages just before civilization (the Bering land bridge is another). Much of the Middle to Upper Paleolithic transition took place along the ice free corridors immediately to the north of this range, whereas the transition to agriculture first occurred along its southern flanks (modified after Figure 17.30 in Dott and Batten, 1988).

not for centuries. (With the possible exception of the general area around the Aral sea, one could look in vain for such a southern passage during the height of glaciation.) In the meantime, human populations could intermittently enjoy prolonged periods of relative freedom from predation pressures.

Indeed, the rarity of large carnivores and the relative abundance of bird and fish bones (indicating fishing and fowling, thus, individual ranging) are characteristic aspects of the Upper Paleolithic archeological sites (Klein, 1989). Of special interest is the decline in the presence of hyenas and lions with the approach of maximum glaciation and in its aftermath (see Table 11.1). These two species of large carnivores represent the primary source of predation pressures on humans. In addition, the depiction of large carnivores in prehistoric art is rare, suggesting their relatively limited distribution in the northern ice-free corridors. At least within these corridors, it seems that humans could now more freely operate as individuals, or in groups oriented only by the task at hand rather than as a safeguard against predation.

The implication is that human agents broke free from the grip of close group formations, which previously were indispensable, but now served

Table 11.1: *Representation of the main carnivore species in fauna/collections associated with paleolithic assemblages from southern Germany (in percentage points)[a]*

Species	Middle Paleolithic 75,000–35,000 BP	Early Upper Paleolithic 35,000–20,000 BP	Late Upper Paleolithic 16,000–10,000 BP
Cave & brown bear	77	94	44
Lion	47	45	4
Hyena	67	30	10
Wolf	67	70	27
Red fox	47	75	27
Arctic fox	35	60	34
No. of levels	49	20	47

[a] The percentage indicates the frequency with which a particular species has been identified from the total number of archaeological levels dated respectively to each of the three time periods (after Gamble, 1978).

no purpose. It is this escape from the confinement of the regimented group that might have ushered action and interaction in a free economic sphere. We know that the opening of a similar window to free individual action (under quite different conditions, of course) was a catalyst for the major innovations in trade, the corporation and other economic institutions, during the Industrial Revolution. The major innovation associated with the Upper Paleolithic "privatization," so to speak, was quite likely the introduction of money.

The monetarization of the exchange system, which I already linked to symbolic behavior, can now be traced to a core area largely confined to the ice-free northern corridors. If we accept this linkage, we would expect to see a relatively strong geographical and temporal concentration of symbolic activity in and around the affected core areas, and much less intense symbolic activity elsewhere (or in the same places at other times). In a way, this is what we observe. Archeologists have long noticed that Upper Paleolithic artifacts identified as symbolic expressions and art objects are largely restricted to northern latitudes. They typically occur along a fairly narrow stretch of land roughly parallel to the maximum glaciation line ranging from southwestern Europe to eastcentral Europe only to reappear some 5000 km (3000 miles) eastward in Siberia toward the northern tip of Asia (Belfer-Cohen, 1988; Wobst, 1990; Bar-Yosef, 1994). In other words, they occur precisely in those parts of the ice-free northern corridor least accessible to migration of wildlife from the south – and thus, least vulnerable to rapid replenishment of locally extinct populations of large carnivores. This is particularly clear in the distribution of the Upper Paleolithic sites that have provided female figurines (see Figure 11.4) and the fact that, during the period in question, art objects rarely occur outside the European world. Indeed, non-utilitarian material is generally absent, as noted by Gamble and Soffer (1990), in contemporaneous sites throughout India, China, Japan, most of Africa, Australia, and the New World. Some authors have even suggested a possible latitudinal component that implies differences in social complexity between western Eurasia and the rest of the Old World around 20,000–15,000 years BP (Lindly and Clark, 1990). This chapter provides an alternative interpretation.

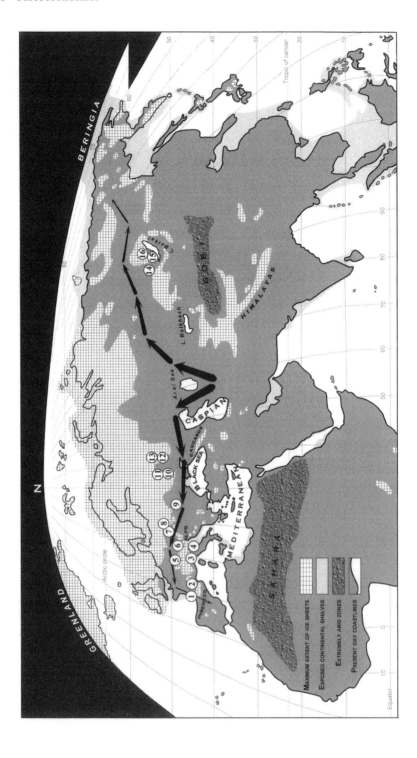

Figure 11.4 **The geography of symbolism, glaciation, and wildlife migration** Upper Paleolithic sites that have provided female figurines or engravings occur largely along a narrow band at the northmost range of the humans distribution during glaciation: the ice-free corridors of western and central Europe as well as Siberia. Most sites are dated to a period concurrent with intensification in glaciation (roughly 28,000 to 18,000 years ago) as mushrooming ice sheets increasingly blocked passage through the Alpine-Himalayan orogenic belt of mountains (see Figure 11.3). The sites are most heavily concentrated in areas that seem to be least accessible to migration of wildlife (especially large carnivores) from the south. Depicting possible routes of migration (on land), each arrow represents 1000 km (622 miles) in length and is proportionally diminishing in width with the cumulative distance from southern origins of migration (that is with the spell of time it takes a locally extinct species to replenish its northern population from its surviving southern stock). Sites marked in the map include (1) Lespugue, Brassempouy and other Pyrenean sites; (2) Laussel; (3) Grimaldi; (4) Chiozza and Savignano; (5) Peterstls; (6) Willendorf; (7) Dolni Vestonice; (8) Peterkovice; (9) Molodovo; (10) Khotylevo; (11) Gagarino; (12) Avdeyevo; (13) Kostienki; (14) Maininskaya; (15) Mal'ta; and (16) Buret.

12 Transition to agriculture: the limiting factor

Against all odds, agriculture was not part of the "creative explosion." The Upper Paleolithic people, completely obsessed with wildlife – especially with grazing animals (as evident, for instance, from their art) – strangely enough found no interest in domesticating any of them.[1] Food production would have been, it seems, the logical conclusion of the Middle to Upper Paleolithic transition, yet for no immediately apparent reason it was never attempted. This failure to act defied for a long time any explanation, although such an explanation is key to any attempt to understand the transition to agriculture, if not to civilization itself.

In 1993 a promising clue was discovered in unexpected quarters. Of all places, it was extracted from a deep hole in the ice-cap over a remote mountaintop thousands of miles away from the nearest site that could conceivably have served as a cradle of early agriculture. It came in the form of a three-page report published that year in *Nature* (364: 218–220) by a group of scientists from the Ice-core Project Summit (GRIP) of Greenland (a country with a land area 99% unfit for agriculture). I will come back to this report shortly, but first we have to establish the necessary environmental and economic connections.

Five unexplained remarkable facts

The first humans who relied on agriculture left more traces in the archaeological record than any culture that preceded them; so much so, that they are a mystery to us primarily because of what we know about them, rather than what we don't. What we know are the following five remarkable facts:

---

[1] Indeed, it seems odd that the most accomplished society of hunters in human history never cared to have a hunting dog. Equally odd is the fact that traders among them kept hauling heavy loads (e.g., of lithic raw materials) over hundreds of kilometers without making the slightest effort to tame a draft animal.

(1) Systematic food production (cultivation of crops and husbandry of livestock) commenced only around 10,000 years ago and spread throughout the world within a relatively short period of time.

(2) Food production was critical neither to human survival nor to human evolution. Anatomically modern humans managed to survive through 90% of their existence in the record without relying on agriculture, and as a whole, the genus *Homo* managed to survive without agriculture for more than 99% of its record.

(3) Agriculture had multiple, apparently independent, origins in widely separated parts of the world at approximately the same time (around 10,000 years ago in the Near East, 8,000 years ago in China, and 6,000 years ago in the New World).

(4) Plants under cultivation and animals under domestication widely differed across different points of origin (wheat, barley and pulses, goats, sheep and cattle in the Near East; rice and millet, water buffalo, pigs and chickens in south and east Asia; maize, beans and potatoes, llamas and guinea-pigs in the New World – to name but some of the most important).

(5) Despite ever improving breeding practices, no new major species (crop or livestock) was domesticated since well before 4,000 years ago; so much so, that in terms of domestication the transformation to agriculture was practically over long before the invention of writing. Indeed, the main techniques of pre-industrial farming (plowing, fertilization, fallowing, and irrigation) were already in common use by preliterate late stone-age people.

Taken together, these facts imply the action of a compelling global, rather than regional, agent. Physical, biotic, or cultural driving forces are three conceivable sources. However, given the geographic separation at the time (especially between the Old and New World) any simultaneous action on the part of biotic or cultural agents can be reasonably ruled out. The global nature of the phenomenon leaves, therefore, little choice but to look for some global change in the physical environment. Because the transition to agriculture was unique to humans, it is clear that such a global change can (at best) provide a necessary condition, but not a sufficient one.

The history of the problem

The prehistorian V. Gordon Childe (1951a,b) was the first to spell out a comprehensive account of the birth of agriculture based on the most obvious contemporaneous shift in the physical environment: the end of the Ice Age. Indeed, the death of the Ice Age and the birth of agriculture seem to share the same approximate date: circa 11,000 BP. Almost a quarter of the earth's land surface had been covered prior to that time by thick ice sheets. At the same time – and not independently – about half the land between the tropics had been occupied by deserts because the water locked up in the ice sheets was unavailable to fall as rain or fill the oceans to their present-day level (of 100 meters and more above the most recent glacial period). Childe was not privy to much of these and other crucial paleoclimatic data which became available only later. His main explanation for domestication, known as the *oasis theory*, was based on aridity-driven symbiosis between humans and animals that, presumably, were forced to share ever diminishing sources of water in the post-Pleistocene harsh environment of the Near East and elsewhere. Childe's general approach had a large following and, as a matter of principle, some consider it sound even today.

However, the concrete details of Childe's model were soon rejected in view of field work evidence obtained, notably, by Robert Braidwood (from excavations in northern Iraq and from other sources). Robert Braidwood and his team of scientists could find little detectible post-glacial effects, along the lines suggested by Childe, on the environment in the region most critical to the emergence of agriculture, the fertile crescent. They soon concluded that swings in climate associated with glaciation had little impact in tropical and sub-tropical regions of the world (save changing sea levels). The single most compelling testimony against Childe's *oasis theory* was the absence of evidence of increased aridity. In their search for evidence of desiccation, Braidwood and others found that the post-glacial period was actually marked by increasing precipitation in the Near East and many other parts of the world. Desiccation in the affected area came into play, we now know, only at the end of the mid-Holocene "climatic optimum" (Bryson 1987). By that time agriculture had been in place some 5,000 years.

However impressive on a plotting chart, and strange as it may sound, changing global averages and long-term trends in climatic variables such as temperature or precipitation may pose little overall threat to the existence of agriculture as a universal phenomenon. Farmers respond to local rather than global conditions and, as a population, they can readily adjust to long trends of climate change simply by converting to new crops or by making geographic moves in a process that, historically, may take place in small increments over many generations. What really concerns a farmer are the small, unpredictable, short-term – year to year and season to season – variations. One day of frost per season is hardly detectable in the annual average but can easily destroy a year's crop. If the event repeats itself only once a decade, the farmer may still thrive. If it repeats itself three or four times a decade, the farmer is out of luck (and business). Going up from one to four unseasonable freezing days per decade implies no global cooling (though a process of global cooling may produce such an effect). An increase in the number of such freezing days can well be produced by a slight increase in day-to-day or year-to-year *variations* even if the *average* temperature is perfectly stable or, actually, increasing over the decades. The same argument can be made about droughts, floods, blizzards, wind storms, "dust bowls," and the like. It is this type of frequent but irregular spiking in weather, if not outright storminess, that helps to wipe out entire local communities of farmers in a matter of a few years or, equivalently, may render agriculture globally unfeasible for hundreds of millennia. This type of unpredictable variation in weather (to which I will return shortly) is precisely what Braidwood and his colleagues could not observe in their data, though such observations were crucial to the issue under their inquiry.

Notwithstanding this difficulty, Braidwood's findings which properly discredited Childe's *oasis theory* helped also to cast doubt on environmental explanations in general. In his own interpretation of the emergence of agriculture, known as the *hilly flank theory*, Braidwood emphasized cultural progress almost to the exclusion of climatic factors. In the following years new studies provided insights gained from new archeological findings, and posited new interpretations: Binford (1968), Cohen (1977), Rindos (1984), Bar-Yosef and Belfer-Cohen (1989) – to name but a few. Overall, the main progress was attained through a better understanding of regional settings and the chronological sequence in each, not to mention the progress in the botanical aspects of agriculture which are beyond the scope of the present discussion. But the question of a global cause for the shift to agriculture, as most authors readily admit, remained poorly explained.

Agriculture versus hunting-gathering

Any explanation for the transition to agriculture is only as good as the distinction it makes between food production and hunting-gathering. There is no shortage of distinguishing manifestations that set the two activities apart. Most apparent among these manifestations are the major tasks that the agents perform (e.g., cultivation versus search and pursuit), the tools they use (e.g., implements versus weapons), the species on which they operate (domesticates versus wild), and the various aspects of lifestyle from habitation (sedentary versus nomadic) to diet (staple versus eclectic). However important from a descriptive point of view, a mere account of such manifestations would tend to fall somewhat short of satisfying our curiosity from an explanatory point of view. In and of themselves they provide little or no clue to the central issue: the strange timing and manner of the shift to agriculture, and the matrix of human incentives in which it took place. In this respect, a slightly more useful approach would be to direct attention not only to apparent manifestations that distinguish the two activities but also to the fundamental procedures underlying each.

Agriculture extended human control of the environment from physical procedures to biotic procedures at the reproductive level of living things. Under agriculture, individual plants and animals were either naturally or artificially selected – i.e., coevolved with humans or deliberately bred by them – for desirable characteristics and for useful products; so much so, it can be argued, that new forms of life are being created on a grand scale. However, at the level of the individual farmer (especially in primitive farming where all tasks must be carried out by hand) the whole question of biotic control of the environment is less heroic, if not ironic. It is true that, unlike hunter-gatherers, farmers control the manner of reproduction of the plants and animals on which they rely. But, from that point on, they themselves fall under the relentless control of their own creation: living things in ceaseless need of feeding, protection, and a host of other services. In this sense, farmers are in the business of following with hard work the life-cycle of animals under domestication and plants under cultivation; so much so, that they have little choice but to run their own daily routines and lifelong plans by the biorhythms of other species. From the viewpoint of the chicken, or the egg, a farmer is scarcely more than an opportune life-support system – antecedent to both.

The temporal nature of farming
As a form of acquisition of natural resources from the biotic environment, agriculture is distinct from hunting-gathering in yet a more fundamental sense. Unlike hunting and unlike gathering and, by extension, unlike fishing or scavenging, farming is first and foremost a temporal activity – regulated and limited by long delays in time – simply because the input-point in agriculture is inescapably distinct in time from the output-point. Almost all crops require relatively large initial expenditures of resources and effort (breaking the ground, sowing, weeding, etc.) long before they yield any fruit. Waiting time is measured by the season and, in the case of orchard-keepers and certain livestock growers, by multiple years. In contrast, the major cost in time in hunting-gathering is the very act of harvesting: search and pursuit, typically, on a circadian time scale. From a purely economic point of view, the prolonged intermittence between the input-point and the output-point in farming is the distinguishing feature that most critically sets it apart from hunting-gathering.

One practical implication is the prevalence in agriculture of sunk cost – cost derived from irrevocable past decisions. Waiting time itself is a form of additional cost: extraneous expenditure entailed merely by the passage of time – unrelated to the agricultural operation itself – but needed in order to make due provision for the farmer's interim subsistence. Agriculture is therefore an activity where the major input of human time and effort (and, possibly, other resources) is made in the form of investment rather than simply in the form of ongoing expenditure. Investment is inherently a risky undertaking. Risk in agriculture is largely measured by variation in weather. Anxiety and apprehension resulting from uncertain future states of nature is more than merely a matter of inconvenience. Unstable climate is an unforgiving limiting factor in agriculture. A society that relies on farming as a *primary* source of living stands no chance of survival in an environment that does not secure a reliable harvest nearly every year.

Risk across states of nature

Nobody can predict a lean year or guarantee a bumper crop. Unpredictable events such as drought, flood, frost, locust infestation – to name but a few – always hang by a hair over the farmer's head. It is true that such natural disasters do not spare hunter-gatherers either, simply because they often affect wildlife as much as cultivated crops. However, there is a fundamental difference in that a hunter-gatherer makes decisions after observing the event, whereas a farmer typically makes the major decisions (and investments) long before observing it. Such unanticipated natural disasters destroy the harvests of both, but the farmer is a double loser, he loses not only the crop, but also the seeds. Consequently, uncertainties about states of nature represent a risk in agriculture that far exceeds the risk in hunting-gathering. Unstable climate makes agriculture an utterly unattractive basis for subsistence compared to hunting-gathering.

With this understanding, it is clear that when we search for a change in climate that could possibly trigger the major shift from hunting-gathering to agriculture, it is not enough to look just at long-term trends in the average levels of temperature and precipitation. One should also, and perhaps primarily, look at annual variation in such variables. When observing ancient climates, attention should be directed to the variance (and perhaps to higher moments of the distribution) rather than to the amicably deceiving means and averages. Finally, it should be noted that beyond and quite apart from the uncertainties of year-to-year variations in weather, there is also the wild card of violent storms capable of destroying not only crops but the entire agricultural infrastructure. The ecological infrastructure on which the hunter-gatherer relies is far less vulnerable to such environmental upheavals.

Climates on average

The multifocal or, at least, dual origins of agriculture (in the Old- and New-World) suggest a triggering shift in some global condition. The only natural condition globally shared by terrestrial organisms occupying

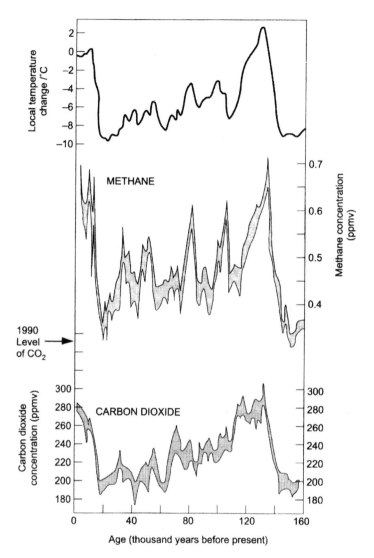

Figure 12.1 **Three indicators of ancient climates** Variation of atmospheric temperature (over Antarctica) and the atmospheric concentrations of carbon dioxide and methane for the last 160,000 years based on observations from the Vostok ice core. The thickness of the lines for the carbon dioxide and methane plots indicates the range of uncertainty in the measurements. As to the variation of global average temperature, it is estimated that it would be on the order of half that in the polar regions (after Houghton, 1997, Fig. 4.4, originally adapted from Raynaud *et al.* 1993, Fig.5).

widely separated parts of the world is the common atmosphere. Indeed, referring to Figure 12.1, there seem to be at least three major shifts either in climate or in the composition of the atmosphere for the period in question: a rapid increase in temperature and equally rapid increases in atmospheric levels of carbon dioxide and methane. A related increase in ocean levels is shown in Figure 11.2. The shifts in all four variables started about 18,000 years ago approaching their peak about 7,000 years later, just in time to coincide with the inception of agriculture.

Topography of heat and moisture
Paradoxically, global warming (or cooling) is to agriculture a regional condition; essentially, a question in geographic location of people and crops. Since temperatures vary widely and quite continuously from the equator to the poles (and from foothills to mountain tops), they leave ample room for temperate zones in between. There is no reason why certain vast areas of the world, at any point in time, will not be thermally suited for agriculture or for any particular crop. Of course, the geographic boundaries of such comfortable zones tend to shift in the long run – say, toward the equator during the height of glaciation or back to higher latitudes in its lull. Such trends may affect the prospects of agriculture in a given region at a particular time, but not the prospects of agriculture as a global phenomenon. Agricultural populations, as I have indicated earlier, are quite able to adjust to *long gradual* trends of a changing climate simply by converting to new crops or making geographic moves over the span of generations, if not the span of a single life. On both counts, adjustment to changing global temperatures (and to many other fluctuations in climate) is more easily reached in mountainous terrain than on a topographically featureless flatland.

Climatically speaking, a mountain operates very much like a refrigerator. With lower atmospheric pressure at higher elevations, air expands as it moves up the slopes and grows cooler very much by the same basic cooling mechanism known from artificial refrigeration (i.e., by expansion of previously compressed gas). For instance, riding an open elevator up a high-rise building, a passenger equipped with a thermometer should notice a drop of at least 1° F (0.6° C) for every 30 floors or so (i.e., about 100 m of altitude) – assuming the thermometer is sensitive enough. The phenomenon is more vividly displayed on the instrument that monitors temperature outside an aircraft during landing or takeoff. Topography permitting, adjustment to climate change through close range relocation in altitude can save long expensive cross-country, if not intercontinental, moves in latitude. The importance of topographically rich terrains to the development of early agriculture has been emphasized by many authors. Noting that in both the Old World and the New World, agriculture first developed in mountainous uplands, Claiborne (1970), for instance, attributes this pattern to the ease of adjustment to climate afforded by the diversity (at close range) in climate itself. That is, to the fact that changing temperatures via altitude operate to compress a whole series of climatic zones into a very small compass:

> A mountain in the Equatorial Zone, for instance, might have rain forest at its base; farther up one might find deciduous trees comparable to those in the Stormy (temperate) Zone. Still farther would come a Subpolar coniferous forest, followed by an alpine meadow that is a fair imitation of the Polar tundra, with an "ice cap" of perpetual snow (and perhaps even a glacier or two) over all. Thus in a climb of perhaps a dozen miles, one can experience climates which at sea level would stretch over some 4,000 miles. (1970:239).

At any time in the course of human evolution there should have been ample spatial variation to compensate for unfavorable long-term changes in global average temperatures – if temperature, as such, was indeed the limiting factor in the development of agriculture. Though nobody can deny the coincidence of agriculture with the trend of global warming in the wake of glaciation, any attempt to draw from it a simple cause and effect environmental explanation for the rise of agriculture should be taken with a large grain of salt.

The same doubts apply to the attempt at drawing similar inferences from rising levels of global precipitation. The trend is evident both, in rising sea levels (see Figure 11.2), and in the changing composition of the atmosphere. In the atmosphere (Figure 12.1) this trend in precipitation is reflected, albeit only indirectly, in the rapid increase in the atmospheric concentration of methane (roughly from 18,000 to 11,000 years ago). I am aware of no direct impact of methane on agricultural crops but, for reasons discussed earlier (Chapter 8), the atmospheric concentration of this substance is a fairly good indication for the worldwide existence of wetlands and, by implication, for the overall intensity of precipitation. Precipitation is another example of a climatic variable which is by no means everywhere alike and, in this sense, its changing global average intensities are again somewhat misleading. Like temperature, rainfall is unevenly distributed in space and time. Furthermore, the resulting moisture actually available to agriculture varies with the topography within each of the world's drainage basins. In response to falling water tables, hill farmers may choose to resettle lower in moister foothills and, if necessary, descend to river banks, changing farming practices from reliance on rainfall to irrigation.

In a way, this scenario led to, or at least was experienced by, the four primary civilizations of the Old World: those of Mesopotamia, Ancient Egypt, the Indus Valley, and northern China. The emergence of all four within a few centuries of each other (roughly between 5,500 to 3,800 BP), but thousands of kilometers apart, was closely associated with the mass-conversion of farming practices from free-range easy reliance on rainfall to concentrated cultivation of alluvial soils based on painstaking irrigation along the banks or derivative canals of major rivers (the Tigris and Euphrates, the Nile, the Indus, and the Yellow River, respectively). Each case in its own way was, apparently, a response to the same lingering climatic event: the worldwide decline in precipitation at the end of the mid-Holocene climatic optimum (to be discussed shortly). Taken together, they bear testimony to the fact that the global *average* intensity in precipitation has implications for the existence of agriculture in a region, but, by the very nature of averaging, not for its global existence.

It is true that much like the associated trend of rising global temperatures, the worldwide post-glacial increase in the amount of water available to fall as rain vastly increased the arable areas of the world, but a shortage in arable land was hardly a problem prior to and during the first stages of the transition to agriculture. In this sense neither temperature nor precipitation could have possibly served as primary limiting (or facilitating) global factors in this transition.

The atmospheric fertilizer
A slightly more compelling avenue of environmental reasoning about the origins of agriculture is provided by the rapidly rising concentration of atmospheric carbon dioxide (CO_2) coinciding (as depicted in Figure

12.1) with the post-glacial process of global warming and jointly peaking with it at the approach of agriculture. Carbon dioxide (in combination with water and light) is needed in order to photosynthesize carbohydrates – the main product of green plants and, in fact, agriculture. Doubling the concentration of carbon dioxide, under controlled conditions, raises by up to 25% the yield of wheat and rice, and by up to 40% the yield of soybeans (Houghton, 1997). In this sense, atmospheric carbon dioxide – which unlike temperatures and precipitations has a perfectly uniform distribution over the planet – seems to be an undisputed universal fertilizer. As such, carbon dioxide enrichment can be viewed as a fairly decisive factor in the global prospects of agriculture – past and future. Could it have been the primary environmental driving force behind the transition to agriculture?

The main difficulty with an explanation for the rise of agriculture based on carbon dioxide enrichment is threefold. First, there is a problem of comparative (as distinct from absolute) advantage. Since the benefits to agricultural ecosystems from CO_2 enrichment are largely shared by natural ecosystems, agriculturalists gain little or no advantage from it over hunter-gatherers. Second, there are also some doubts about the absolute advantage: the size of gains in crops (or in wild vegetation) due to extra CO_2 in the atmosphere. Our information regarding the yield of plants in response to CO_2 enrichment comes primarily from controlled experiments under near optimum conditions. However, when plants grow in a field situation, the optimum conditions required for full realization of the benefits from increased atmospheric CO_2 concentrations are seldom, if ever, maintained. Consequently, the actual increase in yield is far less impressive (Wolfe and Erickson, 1993). Before we get to the third point it is useful to note that the biochemistry of photosynthesis differs among plant species in a way that greatly affects their relative response to carbon dioxide. Most crop and weed species can be classified as either a C3 or C4 type (i.e., species whose photosynthetic products are compounds with 3 carbon and 4 carbon atoms, respectively). The C3 species show significant gains in net photosynthesis from increased CO_2, whereas C4 plants show much less response in this regard. For instance, doubling the concentration of CO_2 under controlled conditions was reported to raise the yield of crops an average 33% for the former but only 10% for the latter (Wolfe and Erickson, ibid.). With this understanding, we can empirically test the following theoretical expectation.

Had the so-called "CO_2 fertilizer effect" been the primary force driving the transition to agriculture, one could expect to see a clear pattern in early plant domestication and cultivation: heavy concentrations of C3 plants almost to the exclusion of their C4 counterparts. In reality, agriculture has never patterned itself on this theoretical expectation. It is true that most of the world's major staple crops (including wheat, rice, soybean, and most horticultural species) are C3 plants that played important early roles in the transition to agriculture. However, nearly as important to agriculture, past and present, are crops that happen to fall under the C4 classification (e.g., maize, sorghum, sugarcane, millet, and many pasture species). The weight of the evidence seems thus to cast some doubts on the possibility that the impact of the CO_2 fertilizer effect, however important, could far exceed the role of a secondary force in the transition to agriculture. For the primary driving force we have to look elsewhere (starting with Greenland).

Climates at variance: a clue in the ice caps

The ice caps of Greenland and Antarctica are made up of compacted snow that accumulated over hundreds of thousands of years in well-defined layers that in certain locations exceed 3 km in depth (the oldest layers at the very bottom are eventually compressed and distorted under the weight above and the ice flows outwards, making room for new layers, in an endless process). Each layer contains valuable information about the prevailing atmosphere and climate going back to the year in which the snow was deposited. To analyze the available information, investigators obtain ice cores drilled from a depth of over 2.5 km (1.6 miles), and thus reach ice formations that fell (as snow) up to a quarter million years ago. Information about the concentrations of carbon dioxide and methane are inferred from ancient air bubbles trapped in the ice. Storminess and other disturbances in the atmosphere (e.g., due to volcanic activity) are inferred from dust particles similarly trapped in the ice. Finally, average temperatures are inferred from the isotope composition of the ice itself.

The quality of information obtained from ice cores is in large measure a question of resolution; that is, the temporal fineness and detail that can be discerned across the layers. The resolution, in turn, depends on two factors: the depth of the layers and the amount of snow deposited in the corresponding years. As one moves down a core to more and more compacted layers and, consequently, as the length of sections representing fixed intervals in time get shorter and shorter, the resolution progressively declines from years and decades to centuries and millennia, and beyond. The length of each section also depends, of course, on the rate of accumulation of snow in each drilling site. At the summit of the Greenland ice cap, the rate of accumulation of snow was, apparently, higher than at the drilling locations in Antarctica (Houghton, 1997). Consequently, the resolution achieved by the Greenland (GRIP) data is sufficiently fine to measure variation in climate on a time scale of a human generation – that is, on the time scale of behavior – going back well into the last glaciation. Moreover, it is sufficiently detailed to detect relatively short-term variations, and *changes* in variation. Some abrupt changes in variation detected in these data were unknown, and unexpected, beforehand. One can hardly underestimate the importance of this piece of paleoclimatic information to the study of human evolution and, in my opinion, it puts the sum and substance of the field in a new perspective. But, to the study of ancient agriculture, it is most important of all.

Figure 12.2 was reconstructed from the GRIP data. It represents a time plot of near-arctic temperatures (deduced from oxygen isotope $\delta^{18}O$ concentration in the ice cores) going back 250,000 years. Note the fine resolution of the data in comparison with Figure 12.1 (reconstructed from the Vostok station in Antarctica). The key finding in this figure – if only from the viewpoint of agriculture – can be appreciated from the summary of the original report (Dansgaard *et al.*, 1993):

> We find that climate instability was not confined to the last glaciation, but appears also to have been marked during the last interglacial . . . and during the previous *Saale-Holstein* glacial cycle. This is in contrast with the extreme stability of the *Holocene*, suggesting that recent climate stability may be the exception rather than the rule.

The *Holocene*, the interglacial of our present time and, roughly, the age of agriculture, is represented in the left panel of Figure 12.2. By comparison

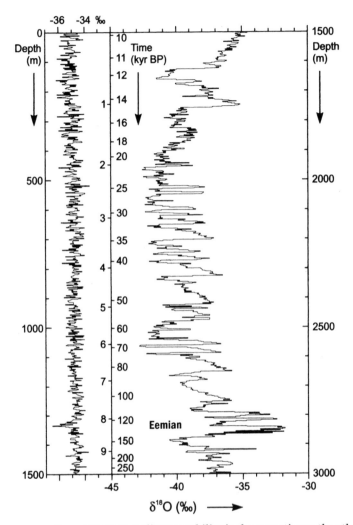

Figure 12.2 **Recent climate stability is the exception rather than the rule** The continuous (3 km deep) GRIP summit ice core is plotted in two sections on a linear depth scale. From surface to 1.5 km depth, the section on the left covers the $\delta^{18}O$ record for the past 10,000 years, From 1.5 to 3 km depth, the section on the right is exceedingly more compressed and, as such, extends the record back 250,000 years. Warmer periods are linked to higher concentrations of $\delta^{18}O$ (less negative numbers in the plot) because relatively less $\delta^{16}O$ is being sequestered from the oceans into the ice. The time scale (in the middle of the plot) was obtained by counting annual layers back to 14,500 years BP, and beyond that by ice flow modelling (from Dansgaard *et al.*, 1993).

with earlier periods, as depicted in the right panel (the panels were deduced from two consecutive ice cores, each 1.5 km long, and are best compared side by side), it is clear that the last 10,000 years have been unusually stable, at least in terms of variation in temperature. Additional data based on deposits of dust particles (not shown here) suggest that the period in question was also unusually benign in terms of storminess. Changes in global temperatures up to 6 degrees Celsius (that is, double digits in Fahrenheit) over periods of several decades are almost unimaginable in the stable environmental setting of the *Holocene*. Yet, changes of this magnitude have occasionally taken place during the upper-paleolithic period (say, 40,000 to 12,000 years ago). Subject to such abrupt swings of climate, to which we have to add the turbulence of unpredictable frequent storms, an attempt to engage in any form of stable agriculture probably would have been abandoned in short order. There was no alternative to hunting-gathering simply because, as we saw, there is no form of agriculture that makes economic sense but one that is stable. Food production as practiced at the moment by humans is in large part still at the mercy of climates and, in this sense, incidently, we are vulnerable far more than other species that happen to practice food production akin to agriculture (Box 12.1).

Box 12.1 Farming in nonhuman societies
Tropical leaf-cutting ants cultivate in their nests fungi on which they feed. A casual observer may wonder how a queen who sets out to start a new nest gets the "seeds" necessary to start a garden in a new destination. During her nuptial flight, the queen, equipped with a special pocket on her head, carries a vegetative pellet of the fungus from her natal nest to the new one. The flight serves the reproductive needs of the ant as well as the fungus. Agriculture differs from straightforward symbiosis in that it consists of the manipulation of the reproductive cycle of one species by another, at the level of individual organisms. In this sense, as I have already indicated (in Chapter 2), food production akin to agriculture in almost all its forms, except for mechanization, can be observed in certain species of insects. This activity has been continuously practiced by them for tens if not hundreds of millions of years: throughout and long before

all recent cycles of glaciation. Recent studies (e.g., by Ulrich Mueller at Cornell University) using ribosomal DNA analysis show, for instance, that the fungal lineage associated with the leaf-cutting ant was "domesticated" at least 23 million years ago.

Farming among the insects is relatively independent of climate variation simply because the activity is primarily, or at least partly, performed "indoors." Crops are sheltered from the physical elements in the internal environment of well-built nests that, for that matter, serve as ideal hothouses – save the benefits of transparency to sunlight afforded by glass. Fungi, a division of simple plants that do not rely on photosynthesis, are a favorite crop among farming ants and termites precisely because they readily grow in the dark when provided with adequate substrate (e.g., cut green leaves). Farming insects use their nests not only as (dark) hothouses but also as sorts of stables. An interesting case in this regard is displayed by a certain species of dairying (or pastoral) ants. These ants keep and tend aphids as a source of food. They milk the aphids (through gentle stroking) for a kind of protein-rich honeydew meal. In return, they provide the aphids protection not only against predators but also from the elements: at the approach of winter they carry their aphid livestock to the nest and in the spring they return them to outdoor plants for "pasture." Whatever the case, risks across states of nature are averted by these and other proven strategies.

By comparison, ours is environmentally a far more opportunistic, if not fatalistic, form of agriculture. Relying on sunlight, human farming is essentially an open-air activity. It has always been dependent in large measure on cooperation of predictable climates. Anatomically modern humans walked the earth for more than 100,000 years awaiting such cooperation. Only at the outset of the Holocene, with the amicable configuration of the present interglacial climate, could they for the first time seriously embark on food production.

Agriculture was not a practical option for the upper-paleolithic people, nor for their forebears going back at least to the previous glaciation and, more intriguingly, to the previous interglacial – the *Eemian*. The *Eemian* (roughly, 125,000–115,000 BP) was at least as long and as warm as the present interglacial and, in that sense, may raise a thorny question. Noting that anatomically modern humans, quite likely, were already walking the earth during Eemian times, the question is why did they fail to initiate some form of agricultural production? Did they suffer from some inhibiting defect in rational behavior, unobserved in their anatomy, that prevented them from taking full advantage of economic opportunities? The GRIP data seem to settle the issue. It is evident from Figure 12.2 that the *Eemian* interglacial, however warm, was no less volatile than the two glaciations preceding and following it and in this respect could not have been any more hospitable to agriculture.

In sum, it is clear that one of the major features that distinguishes the climate of the last 10,000 years or so from earlier climates is stability. This major finding from the Greenland ice cores data adds a remarkable new dimension to our understanding of the conditions under which human prehistory evolved. Without this finding it is difficult to point to any single major environmental cause for the shift from hunting-gathering to agriculture. It provides what global averages in major variables (temperature, precipitation, and atmospheric carbon dioxide) fail to provide; namely, a compelling *necessary* condition for that transition – but not quite yet, a *sufficient* condition.

The Fertile Crescent: a regional case study

In addition to the global issues associated with the transition to agriculture there are many regional questions. Chief among them is the question of the birthplace (or places) of agriculture. This question draws much attention to a relatively small area in Southwest Asia known as the Fertile Crescent. We do not know if this (or any other) region can sensibly be described as the cradle of agriculture, but a rich body of evidence of cultivation and domestication at their earliest known dates (10,000–8,000 BP) occurred in more than 50 major archeological sites throughout the region and provides at least a partial justification for the label.

Under its current climate, the Fertile Crescent is impressively fertile relative to the barren deserts surrounding it, but not necessarily relative to more distant places. Many regions in various part of the world, from sub-Saharan Africa to western Europe to the Americas, harbor equally favorable or superior combinations of soil, moisture, and sunlight. The exceedingly fertile San Fernando Valley of California which lies roughly at the same latitude a globe apart from the Fertile Crescent is a case in point – and there is no shortage of other examples. For more than 5 millennia before the advent of agriculture, people were present on all continents (with the exception of Antarctica) and could, no doubt, gain access to many optimal areas. In certain cases pre-agricultural people were probably already walking those grounds in their capacity as hunter-gatherers. This being the case, no one can leave the ancient farmers who first embarked on full-scale reliance on agriculture in the Fertile Crescent without asking a simple question: what was agriculturally so special about this particular place.

These questions can be approached, and perhaps partly answered, by noting that prehistoric farmers in the region (and elsewhere) probably enjoyed a climate more benign than their historical counterparts – present-day farmers, included. With some caveats and uncertainties (e.g., Bryson, 1987, Hu et al., 1999), climate studies generally show that the "climatic optimum" of the present interglacial occurred within its first half – roughly between 8,500 to 5,000 BP – with conditions deteriorating throughout the second. The deterioration is evident in cooling average global temperatures, falling water tables, disappearance of freshwater lakes and reappearance of drifting dune fields, as well as in a decline in rainfall almost everywhere around the world. In the general area associated with the Fertile Crescent this trend led to endemic droughts and expanding deserts of the kind that drove the ancient Akkadian civilization to extinction (around 4,200 BP) and placed Jericho, one of the earliest known sites of agriculture, into its present-day near-desert setting. I do not know if the Garden of Eden is a proper metaphor for the amenities afforded by nature 2 millennia and more before the bible, but there is little doubt that the climate at the time was far more salubrious than anytime thereafter. We can clearly conclude that the entire agricultural enterprise in the region was climatically in a better position during its formative age than it is today.

Whether the Fertile Crescent was also in a position of *comparative advantage* against the rest of the world, is a slightly more complex issue. The climatic optimum of the present interglacial was not confined to any particular region of the world, nor was its subsequent deterioration.

For instance, it was probably no coincidence that the desiccation of the Middle East occurred in close juxtaposition with the retreat of treelines as far away as northern Canada. Likewise, at the cultural level, it is probably no coincidence that the fate of the Indus civilization followed in fairly short order (around 3,600 BP) the demise of its Akkadian counterpart, apparently for much the same reason: lack of rainfall. Evidently, global trends are not always the most compelling source of regional explanations for local comparative advantages (or disadvantages). Fortunately, there exists a truly regional alternative explanation based on the unique geomorphic setting within which the general area of the Fertile Crescent – the Near East – is situated.

The Fertile Crescent is surrounded by five major bodies of waters: the Mediterranean to the west, the Caspian and Black seas to the north, the Persian Gulf to the east, the Red Sea to the south – all adjacent or barely 300 km (186 miles) away. Each piece in this hydrographic mosaic is large enough to have a moderating effect on the atmosphere in its vicinity (and in the case of the Mediterranean even far beyond). Taken together, they probably have a great moderating effect on the entire region they so closely encompass. Standing alone, each is nearly, if not fully, a landlocked pool of water too small to develop or entertain the kind of ocean circulation (i.e., cool and warm surface drifts and currents) that make the weather so unpredictable in many parts along the Atlantic and especially, the Pacific coasts. In fact, checking the global position of the Fertile Crescent, one can scarcely find a spot of land further removed (simultaneously) from the shorelines of the Atlantic and Pacific oceans, with the Indian Ocean 1,500 km (1,000 miles) away. I don't know how all these factors can be put to play in a rigorous model of regional climate, but the consequences speak for themselves. Windstorms anywhere near hurricane force and devastating widespread floods, as evident from Table 12.1, are practically unknown in this part of the world. It is true that the near eastern climate is subject to sharp season-to-season and place-to-place variations. But these, by their very nature, are highly predictable. The unpredictable components of the climate – i.e., day-to-day and year-to-year variations – are at the same time only minimal compared with other regions of the world. In other words, the atmosphere over the Fertile Crescent offers ample diversity but little in the way of surprise. This configuration of predictability combined with environ-

Table 12.1. *Notable[a] storms of the 20th century by region and casualty count*

Region of the world	Windstorms and blizzards		Floods[b] and tidal waves		Total	
	Events	Deaths	Events	Deaths	Events	Deaths
Mainland Asia (Near-East excluded)	27	593,510+	29	1,046,121+	56	1,639,631+
North America	34	14,166	24	9,741+	58	23,907+
Australia and Pacific Islands	19	12,729+	6	8,664	25	21,393+
Caribbean and Atlantic Islands	21	14,093	0	0	21	14,093
South and Central America	3	2,281	12	7,018	15	9,299
Europe	1	1,000	11	4,177+	12	5,177+
Africa (Egypt excluded)	0	0	5	2,188+	5	2,188+
Near East[c]	0	0	1	19	1	19
Total	105	637,779+	88	1,077,928+	193	1,715,707+

[a] As listed in the *World Almanac and Book of Facts 1999* (New Jersey: Primedia Co.).
[b] Excluding dam collapses.
[c] Including Turkey, Lebanon, Israel, Jordan, Syria, Egypt, Iraq, Saudi Arabia and the other countries of the Arabian Peninsula.

mental opportunities for diversification in crops and livestock is, on both counts, favorable to farming. In the formative stages of agriculture it was, apparently, more important than ever.

On the whole, the global picture one can draw from the data in Table 12.1 is probably pertinent to the relatively favorable conditions (in terms of climate stability) during the current interglacial period. On the other hand, it tends no doubt to trivialize the unfavorable conditions at the approach of the climatic optimum, just prior to agriculture. We know that violent swings in weather typical of glaciation continued to destabilize the global environment long after the ice sheets started their final retreat (see Figure 12.2). The last of these major swings in weather (known as the *Younger Drays*) came to an abrupt end only around 10,700 years ago. Until that time the climate even over the Fertile Crescent was far more volatile (and unpredictable) than it is today. The main point, however, is that the climate almost everywhere in the rest of the world, quite likely, was contemporaneously even more volatile. In fact, the fundamental geomorphic conditions providing relative stability to this part of the world were apparently in place long before all the cycles of recent glaciations (going back perhaps 5 million years to the time when the Mediterranean was reconnected to the Atlantic ocean and, consequently, was filled up to its present level). Since then, speaking strictly in comparative terms, the area in question was probably always an island of *relative* calm in a world of climatic upheaval. With global swings in weather subsiding in the wake of glaciation, it is reasonable to assume that the prevailing environmental conditions stood the best chance of reaching a critical point favorable to agriculture in this region, of all regions, first. Indeed, the fact that full-scale human reliance on cultivated food occurred throughout the Fertile Crescent fairly soon (i.e., no more than a few centuries) after the *Younger Drays*, seems hardly to be a coincidence, either in space or in time.

In sum, examined in juxtaposition with the corresponding paleoclimatic record, the archeological record of the Fertile Crescent seems to lend further support to one of the fundamental economic arguments in the discussion so far; namely, that risk and uncertainty across states of nature is a paramount constraint in agriculture. Climate instability, more than any other environmental variable that meets the eye, was the limiting factor that blocked the transition to agriculture for an exceedingly long period of time in antiquity. Yet, the relaxation of this constraint in the wake of glaciation could not, I shall now argue, set the transition into motion without the facility of human exchange.

13   Transition to agriculture: the facilitating factor

The specialization–diversification dichotomy

Our life as productive agents is incredibly narrow in scope. Most people, when asked to list the types of products (commodities or services) they help produce in a course of a working day, if not a lifetime, would be hard pressed to point to more than one item. On the other hand, with the exception of the most devoted ascetics, almost everybody can easily list a dozen or more distinct products they help to consume before breakfast is over, and dozens more by the end of the day. Strictly in scope, though not necessarily in craftiness or subtlety, our life as consumers seems to be exceedingly richer than our life as producers.

The pattern is of great antiquity – as old, in fact, as hunting-gathering (when the act of procurement was first set apart from the act of consumption). This state of affairs reflects a fundamental dichotomy in the principles of human production and consumption. In production, the central organizing principle is *specialization*. In consumption, quite the opposite, the central organizing principle is *diversification*. By their very nature, these two principles are at constant variance with each other. The conflict can be reconciled only by redistribution and that is, in the final analysis, the primary function of exchange.

Agriculture rose in the face of this conflict. The transition from one-step acquisition to multi-stage production of food greatly intensified the tension between the expediency of specialization and the necessity of diversification in diet. This worked to put great demand on the system of redistribution and, by extension, on exchange. The redistribution of food in society could no longer rely entirely or primarily on the sexual division of labor and on informal (nepotistic) exchange but, inescapably, became increasingly reliant on the networks of formal trade.

The question of autarky

Pastoral depictions of agrarian life under the ideal of self-sufficiency would have given little comfort to ancient farmers. Their displeasure is expressed in skeletons dating to the time of the first human acquaintance with agriculture. Small overall size, knobbly joints, thinning in the outer layer of bones, abscesses and dental caries – all bear testimony to diets badly out of balance. Considerable loss of diversity in human diets due to specialization in food production is also evident in the material remains of food itself. Thus, for instance, a dramatic decline in the number of edible species of plants in human use has been inferred from traces of seeds and fruits occurring in archeological sites that span the period of transition from hunting-gathering to agriculture (e.g., Abu Hureyra in northern Syria). Self-sufficiency could have made a bad situation only worse. In reality, food production entailed ever-increasing interdependence among individuals and across communities. The transition to agriculture could hardly have been accomplished in the regions of the world most closely associated with it, had it not been accompanied by the intensification of the division of labor and redistribution, and thus exchange, to levels unknown among hunter-gatherers.

For reasons discussed in Chapter 5 (in connection with the expensive tissue hypothesis) humans, perhaps more than any other group of organisms, rely for subsistence on a finely diversified diet. This includes, at first approximation, a balanced intake of carbohydrates (primarily from plant matter) and high-quality protein (from animal matter). Prior to agriculture – that is, for more than 99% of the time the human digestive system was evolving under encephalization – gathering was the main source of carbohydrates and hunting the main source of protein. These two activities could be performed at relatively close (radial) range by a family or a small kin-group, for much of the wild vegetation and sources of fresh water sought by humans and their game typically occurred in the same eco-system. The transition to agriculture changed this old arrangement in a fundamental way. It effectively split the human feeding ecology by separating in time and space the procurement of plant matter from the procurement of animal matter.

Farm animals are part of the scene in almost any farming settlement typical of traditional agriculture: a pig or a handful of unattended chickens scavenging around the living quarters and a single cow or a goat, perhaps a donkey, browsing nearby on low-grade forage. Unlike herd or stable animals, livestock in this category are typically not intended, nor can they possibly provide, for the kind of steady supply of protein needed to maintain a farming family in good health. They are intended, instead, for more narrow purposes. Some may serve as draft animals more likely to die of sickness or old age than reach the dinner table (Banning and Köhler-Rollefson, 1992). Others serve as a living storage instrument that converts inedible organic waste and surplus crops into durable proteins. These are slaughtered and eaten or traded away for other foods only on special occasions, typically, as a last resort in bad times. In other words, they serve as hedge against fluctuations in nature and bad luck – the only insurance policy a farmer traditionally could opt for.

To be truly self-sufficient in adequate provision of high-quality protein from animal matter, if only as an essential supplement to diet for a family of five, it is incumbent on a farmer to maintain a standing herd of about two dozen heads in small stock or the equivalent in cattle. (A herd twice as large is needed if animal matter is to serve also as a major source of energy.)[1] Tending animals on top of the chores of land cultivation would make great additional demands on time and effort, but the limiting factor is the sheer quantity of food a herd of this size must consume day by day. An animal consumes at least 10 calories for each calorie it can possibly provide in protein or fat. Thus, to borrow words from Jared Diamond (1997:169): "it takes around 10,000 pounds corn to grow a 1,000-pound cow." At this rate of energy (or biomass) transformation, and with cultivation carried out by hand, early farmers could ill afford even to imagine the possibility of growing feed crops for the purpose of maintaining livestock. Indeed, the kind of sedentary animal husbandry practiced in modern agriculture (where livestock on many occasions are kept in stables and enclosures and fed accordingly) is not, as a rule, the basic form of economy in primitive and traditional societies (Khazanov, 1984:24). There is simply no option but to herd livestock into pasturable land. Pastoral husbandry, however, has its own distinct constraints that for the most part are incompatible with land cultivation.

In terms of land use, pastoralism is by far the most extensive form of agriculture. Perhaps more than any other traditional (i.e., pre-mecha-

---

[1] This is actually a fairly conservative estimate for the minimum requirements of a family in a full-scale pastoral society. Higher estimates are often listed in the literature (e.g., Khazanov, 1984:31–33).

nized) agricultural activity, it also offers large economies of scale. For instance, contrary to conventional perceptions, it is often more difficult to control a herd of a dozen heads then a herd ten times larger, simply because the herd instinct of many animals is activated only when clustered in large dense masses. Thanks to this instinct, herdsmen around the world are known to cope single-handedly and quite comfortably with herds up to 300–400 head of small stock even in difficult terrain and without the aid of a horse or a dog (ibid.: pp. 31–33). This exceeds by a great margin the needs of a single family. A large herd quickly eats up, of course, all the grass in its immediate surrounding and – especially on marginal land and in more arid regions – must be kept constantly on the move along considerable distances away from any fixed geographic point. It is clear therefore that pastoralism, by its very nature, would tend to reward mobility and penalize sedentarism.

Land cultivation, quite the contrary, tends to reward sedentarism and (beyond a certain point) penalize mobility.[2] Cultivation itself breaking the ground, sowing, weeding, watering, and harvesting is a sequence of tasks performed round the seasons on the same spot, typically, a single plot of land, and this is where farmers spend most of their working time to begin with. Consequently, a farmer's capital is largely invested in sessile crops: a fairly vulnerable asset at all seasons. Starting with the seeds, throughout all stages of cultivation, and eventually in storage – crops are easy targets to pests, wild life, the physical elements and, indeed, to human mischief. Farmers, therefore, can ill afford to stay away from their fields and storehouses. Keeping livestock on pasture under these circumstances is quite an undertaking. Since one person cannot be present at two places at once, that task must be performed by a separate family member or be entrusted to a hired (or otherwise designated) member of the community at large. Farmers, however, do not like to see their capital go unmonitored if only for reasons of good agro-business practices:

---

[2] A certain degree of mobility is essential in all economic activities and land cultivation is no exception. Seasonal mobility associated with fallow crop rotation (in temperate zones) and with slash-and-burn horticulture (in tropical zones), are obvious examples. In the long run, adjustments to changing environmental (or technological) conditions sometimes entail selective or even mass migration: a frequent episode in the annals of agriculture that did not spare modern farmers. John Steinbeck's novel *Grapes of Wrath* bears literary testimony to this effect if only in one historical instance: the dust bowl years, in the central United States during the 1930s.

"The master's eye is the best fertilizer," noted long ago the Roman naturalist Pliny the Elder (23–79 AD).

The existence of farms that are engaged in both cultivation and herdsmen husbandry can be a relatively stable economic and social system only under conditions that permit short-range pastoralism: that is, typically, a situation where livestock driven to pasture in the morning can be routinely recalled by nightfall. This requires the availability of round the year well-watered (i.e., rapidly regenerating) pastures situated very near a farm or a village: a configuration which is usually not easy to come by. But even when such ideal conditions are approximated (e.g., in certain regions of western and central Europe) agro-pastoral mixed farming is practiced not so much at the level of a single farm as at the level of the community at large. Typically, some farms in the community are engaged primarily or exclusively with pastoralism along others that are engaged primarily with cultivation. The community, as a group, is duly diversified though each farm separately might be highly specialized.

Indeed, even under conditions ideal for autarky, an arrangement in which farms in a community tend to be heterogeneous but the tasks performed within each farm are fairly homogeneous is bound to achieve better results in terms of overall output than outright autarky (in which farms are homogeneous but the tasks performed in each are highly heterogeneous). However, this arrangement already requires a degree of redistribution in the form of (local) exchange in food items.

Less than optimal conditions call for a larger extent of redistribution and exchange. Under conditions typical of arid and semi-arid regions of the world, short-range pastoralism is no longer an option. To the extent that rainfall is scant or unevenly distributed throughout the year the regeneration of pastures gets sluggish and intermittent. Herds, if they are to survive under these conditions, must be driven over extended distances and cover vast stretches of land (especially in the dry season, where applicable). An attempt to maintain a mixed farming economy where land cultivation is performed in close juxtaposition with pastoralism would tend, under these conditions, to produce pressures that threaten not only the integrity of the family but also the structure of society itself. (Interestingly enough, these pressures would differ in degree depending largely on a single, seemingly unrelated, variable: rainfall.) A split that set apart land cultivation from animal husbandry in ownership, location, and lifestyle of the human productive agents – perhaps in culture – is almost inevitable. Equally inevitable, however, will be the need to share somehow the separate final products (i.e., carbohydrates and protein, respectively) among the members of society at large, for everybody needs to meet, in the final analysis, the same basic human requirements of a balanced diet. This process of differentiation in production and recombination in consumption can be achieved, it seems, only under exchange.

At issue is local exchange in food items. Unlike long-distance exchange in durable items (e.g., lithic raw materials or ornaments) for which there is ample evidence in the prehistoric record, local transactions in perishable food items left little or no direct traces in this record. Since experts, understandably, pay far greater attention to evidence in hand than to evidence absent, a casual observer may be left with the impression that ancient farmers or their contemporaries were readily engaged during the formative stage of agriculture in the exchange of obsidian, seashell, and similar non-food durables items – to the exclusion of food itself. In reality, local transactions in food items far exceeded any other form of exchange in almost any economy until recent historical times and the formative years of agriculture were probably no exception. The fact remains, however, that there is little in the way of direct evidence to prove or deny this contention. Fortunately, there are some indirect pieces of evidence that bear on the issue.

One source of indirect evidence for the existence of local exchange in preliterate (and prenumismatic) societies can be drawn from prehistoric configurations of material structures, both organic and inorganic, that have no reasonable explanation without exchange. In what follows I will discuss two such structures as they are associated, in order, with domestication and with early patterns of human settlement. Both, and each in its own way, bear upon the role of exchange in the early stages of the transition to agriculture.

The Caprine paradox

Poorly understood without exchange (and thus indicative of it) is the role of some animals under domestication. Chief among them are the goat and the sheep. Next to the dog (a one of its kind product of preagricultural domestication), the goat and the sheep were the first animals ever to be kept in human company, though, between the two, we do not know with certainty which came first. Whatever the case, the important point is that both, if not simultaneously in nearby regions (of the Fertile Crescent), then in short order one after another, the goat and the sheep were the first animals ever to be domesticated for the purpose of food production. At one level – that is, in their capacity as house animals (distinct from herd animals) – the domestication of the goat and the sheep and, for that matter, the pig, seem perfectly understandable. As I have already indicated, the main function of animals in this capacity is largely confined to recycling of degraded farm products and vegetation into a durable, though secondary, source of protein and valuable by-products. What goats and sheep in this humble category can offer is, at best, a small but steady supply of milk and fibers (perhaps blood) and, once in a while, an opportune source of some extra meat. This might well qualify the goat and the sheep as candidates for early domestication, but it does not quite explain the final role they were actually to play in the history of agriculture – indeed in human history – that, by any standard, far exceeded the role of house animals.

To fully understand the central role of sheep and goats in the emergence of agriculture, it should be recognized that the transition from hunting-gathering took place along two parallel roads: the road from gathering to cultivation, and the road from hunting to pastoralism. Sheep and goats played a pivotal role in the latter. For the first 4,000 years, while the agricultural world was awaiting the domestication of cattle and other large stock, small stock served as a primary source of high quality protein in the diets of agricultural populations. Under human control, moreover, herds of sheep and goats effectively crowded out much of the wild populations of grazing and browsing animals in almost any area of the world that came under human occupation and, by extension, cut down on the presence of large predators by depriving them of their main prey in the affected areas. All the advantages of a predator-free zone that helped unleash the creative forces of the Upper Paleolithic people along the ice-free corridors to the north of the Alpine-Himalayan belt of mountains (Figure 11.3) could now be replayed with added force to the benefit of agricultural people along its southern flanks.

The discussion in this section is primarily intended to show that caprines (as the term refers to sheep and goats combined), were highly suitable and well-timed objects of exchange in a market system with food becoming an increasingly important item in formal trade. As such, they helped expedite the division of labor in food production. Promoting large-scale pastoral nomadism and facilitating urban life, caprines had a profound effect on patterns of human settlement and, by implication, on social and cultural structures. On the down side, the image of the caprines as benign grazers is best taken with a grain of salt. Goats, in particular, are quite capable of stripping their immediate surrounding of much of its non-grass vegetation, thus, accelerating dessication. In that capacity they have actually brought to ruin the ecology over vast areas, notably around the Mediterranean. Finally, the historical significance of the human–caprine relationship – for better or for worse – is evident in the test of time. Today, pastoralism remains the most common mode of human (and caprine) livelihood that has endured almost without change since it was first initiated at the earliest stages of agriculture.

The world's caprine population is currently estimated at roughly 5 billion head – nearly enough to place a sheep or a goat in the company of every man, woman, and child present on earth. With so many sheep and goats in our company we tend, no wonder, to take their presence for granted. We also have ample opportunity to better understand the role of pastoralism past and present. What is more difficult to understand, however, is how pastoralism could have so completely displaced hunting to begin with. Why did humans for the first time, and of all times, choose to rely on domesticated stock precisely when (due to the climatic optimum) wild stock in many parts of the world was more abundant than ever?

Husbandry is a labor-intensive undertaking. It takes, in general, more time and human effort to raise a domesticated animal than to hunt its wild counterpart. In the course of domestication an animal stands to lose (among other expensive structures that are no longer essential for its survival) physical dexterity, intelligence, and the acuity of sense receptors; in fact, its overall ecological bearings. The animal can no longer fend for itself. Unlike its wild counterparts, a domesticated animal cannot on its own find optimal sources of food and water, locate and exploit shelter against the elements, protect itself and its young against predators, or even regulate its own reproductive cycle. All these vital functions are provided by the human handler at great costs in time and effort. One lucky strike with an arrow can earn an expert hunter of gazelles or wild sheep, for instance, the same amount of meat and near-ly all the by-products (skin and fiber, though not milk) that a herder will obtain only by long hours of toil over months if not years in waiting. (Procurement of milk, the only clear advantage of animal husbandry over hunting, plays a primary role in dairy farming but only a secondary, if not marginal, role in free-range pastoralism.)[3] On the face of it, the early and almost complete displacement of hunting by pastoralism makes no more economic sense than, for instance, a similar displace-ment today of all our open water fishing with fish farming. There must have been a fairly compelling reason for such a radical displacement in the primary mode of procurement of animal matter to actually take place. In calling this reason to account, it would be useful to view meat not merely as a trophic item in the diet but also as an economic entity – a commodity in the market.

A commodity is much more than the sum total of its intrinsic physical attributes. Each unit of a commodity has, to begin with, a distinct date and an address – a point of delivery in time and in space – that along with (and sometimes despite of) its physical specifications will deter-mine its final value in human use. An umbrella on a rainy day, for instance, is of greater value to its carrier outdoors than indoors or on a sunny day, whereas a second umbrella of exactly the same physical speci-fications is generally considered a nuisance to carry around under any circumstances. Meat is no exception to this rule. A pound of meat obtained from a wild animal (e.g., a gazelle) and a corresponding pound of meat obtained from a domesticated one (e.g., a goat) are hardly dis-tinguishable in physical specifications even under the most pedantic of inspections, in the laboratory or on the dinner table. Yet under the scrutiny of economic reasoning they represent two widely separated cat-

---

[3] Milk products are believed to be items of secondary importance in the dietary system of prehistoric nomads (Bar-Yosef and Khazanov, 1992:6) as is true, to some extent, even of modern nomads (Khazanov, 1984:39). Consistent with this view is the fact that the cow, our primary milk animal, always played a distant second to sheep and goats in the ancient Fertile Crescent, and indeed, was the last of these three species to be domesti-cated in the region (Smith, 1995:65-7).

egories of commodities. In the absence of refrigeration, the meat obtainable from the gazelle, and game in general, is a highly perishable commodity because it comes into human possession only after the animal is dead. As such, it is unfit for human consumption anywhere out of close range from the kill in space or in time. The implication is that hunting can be an important resource in human subsistence only if the population distributions of humans and wildlife closely coincide; that is, only if points of game acquisition and points of demand for butchered animals occur in close juxtaposition. While such a pattern of human settlement fine tuned to the availability of wild life is commonplace in hunter-gatherer societies, it is no longer possible or desirable under agriculture. In fact, the emergence of food production, in particular specialized food production, would tend to separate in space and time producers from consumers with points of food acquisition and corresponding points of demand drifting apart to ever longer distances. The appearance of embryonic urban centers almost at the earliest possible stage of this transition (to be discussed in the next section) is only one manifestation of this trend. So dominant in the world of hunters-gatherers, game animals were soon bound to become a marginal resource in human subsistence, for their flesh could no longer reach, in edible condition, the vast majority of their eaters.

In contrast, the meat obtainable from the goat, and domesticated stock in general, is a highly durable commodity because it comes into human possession in a live animal. It can stay fresh and perfectly intact in its live storage for years, if not for decades, on end – at all times available for consumption on demand. Unlike their wild counterparts, domesticated animals can be manipulated with great temporal flexibility. Herd animals, in addition, are also highly mobile in space. Herds of sheep and goats can be driven with relative ease along tortuous trails to reach distant pastures in the most difficult of terrains. Thanks to this capability, these herds can also be driven to and across distant markets and, as such, are likely to end up in the market where they command the highest value. Of special importance in an economy yet lacking beasts of burden, to say nothing about refrigerating facilities, was the fact that these animals could deliver themselves to the marketplace, if not to the kitchen, alive and on all fours (i.e., in fresh condition and at great savings in transportation cost, respectively). All these advantages more than justified the extra time and effort involved with pastoralism – assuming, of course, that there was a market for the products of a herd somewhere beyond the immediate community of the herder. In the absence of such a market, hunting would still produce the same or better results at less effort.

In sum, the domestication of sheep and goats at the earliest stages of agriculture, and the almost complete displacement of hunting by pastoralism shortly thereafter, could make little economic (and common) sense had there not been, in time and place, a fairly developed or at least developing system of market exchange in food items.

Agrarian origins of ancient cities

Plato, himself a citizen of a city state, traced (in the *Republic*) the primary reason for the existence of cities to the urban division of labor and exchange. Partly overlooked in this approach was the fact that the fabric of division of labor does not stop at the city walls. Modern urban and regional economists usually pay closer attention to the rural hinterland. Following the agricultural economist Johann Heinrich von Thünen (1783–1850), they trace the origins of the city (at least of the provincial town) not so much to the *urban* division of labor, as to the *agrarian* division of labor.

Thünen's (1966) model of optimal land use around and (to a lesser extent) within a city was based in part on his own farming experience. As a Prussian landowner of some prominence he had ample opportunity to acquire such experience first hand. In large part, however, his model was based on his higher calling as one of the eminent theoretical economists of the Napoleonic era (and the undisputed founding father of modern location theory). The model was constructed around the observations that at least to some extent farm products are objects of exchange; that farmers seek to maximize incomes and, to that end, are willing partly or fully to specialize in crops along optimal lines; that land goes to the highest bidder; and that, in addition to land, transportation is a primary limiting factor. Under these conditions, the price of grain, for instance, would have to be such that the land on which it was produced could not be used more profitably for any other product, agricultural or otherwise.

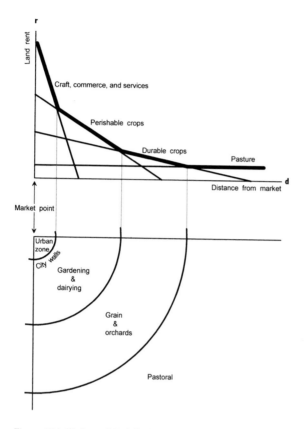

Figure 13.1 **Thünen Model of agricultural land use and the formation of an urban center** An efficient (concentric) pattern of land use emerges as farmers choose crops to maximize income and land goes to the highest bidder. The formation of the central business district (in size, shape, and spatial position) is endogenous to the model unless overriding exogenous forces are at play. The model can be reconstructed from the market equilibrium condition $r = y\,(p{-}c{-}df)$ indicating that at each point the rental price of land (r) is controlled by yield per unit of land (y), market price (p) and production costs (c) per unit of product, as well as the distance from the production point to the market (d), and the unit freight cost per unit of distance (f). Subject to this market condition, Thünen Model generally predicts that at each location a farmer will tend to select the crop that maximizes the value of r at this particular site. Moreover, if all farmers follow the rule, then the total *value* of the agricultural product available to society net of production and transportation costs will be maximized, as well.

Under the same conditions in general, and much in the same manner in almost any conceivable spatial setting, the exchange in farm products is expected to produce concentric market forces that eventually lead to the creation of a fixed market point surrounded by well-defined rings of farmland assigned to the cultivation of distinct classes of crops or livestock, each depending on its distance from the center (see Figure 13.1). Productive activities that take little land space or require close proximity between providers and recipients, or both (e.g., crafts, commerce, and various civil and personal services) would tend to gravitate toward the center eventually forming a densely populated urban or proto-urban center. Conversely, productive activities that require vast areas of land but only infrequent interaction between trading partners (not the least, pastoralism) will tend to recede to the outer ring. In an ancient city setting, this outer ring might extend considerable distances from the center and conceivably include uncultivated wilderness suitable only for nomadic activities. In between, moving from the center to the outermost periphery, cultivation would gradually become more and more extensive over intermediate stages. The inner ring just over the perimeter (or walls) of the urban zone would be largely dominated by intensive cultivation of perishable crops and, in arid and semi-arid regions, might require irrigation. As to livestock, the main concern near the city would be lack of pasture with the proximity to the point of demand providing the compensating balance. Hence, except perhaps for some dairy cattle, the emphasis would be on species that can feed on surplus crops or organic refuse but are otherwise not particularly suited for long travel (e.g., pig, rabbit, and chicken). As the distance from the center lengthens, the emphasis would gradually shift to more durable and less bulky crops (e.g., grain) as well as to livestock that could easily be corralled and chaperoned to the distant market (e.g., goat and sheep), and irrigation would be gradually replaced by reliance on rainfall.

Over the years, Thünen's model has been expanded and modified by economic students of location theory. To fit the changing conditions of rapid industrial urbanization in developing countries, it has been increasingly stripped of its original agrarian context and content. However, the fundamental principles remain largely intact (sometimes under different names) and, as such, have since become the standard against which observed patterns of urban and regional land use are actually tested. It is Thünen's model, largely unmodified (if not precisely in its original rendition) which can provide, in my opinion, the best description yet for the formation and structure of the ancient city in the context of its agricultural hinterland.

Simply because towns and cities (ancient or modern) cannot survive without ongoing import of food, their mere existence implies a certain degree of division of labor and redistribution between the rural and the urban populations and, by extension, a similar degree of ongoing exchange between the parties in the form of *food for non-food items*. On this point, Thünen's model is far more explicit and conclusive. The model implies also a degree of division of labor and redistribution, thus exchange in food items, within the rural community itself (with or without the aid of a central market). As farmers specialize (within each of Thünen's rings of optimal cultivation), they become less and less self-sufficient and more and more inclined to trade with their (close or distant) neighbors that operate on different rings and cultivate different crops. Specialization within the Thünen's rings is both supported and supports exchange across such rings. Of course, exchange among farmers can, but need not, be conducted on the spot (or over the fence) on a one-to-one direct basis. It is usually less awkward and more economical (in terms of transactions cost) to break up transactions into two or more steps through the use of money or intermediaries and conduct business in the central marketplace, which partly exists precisely for that purpose. Whatever the case, the important point is that in the final analysis exchange of *food-for-food* is affected. The implication is that specialization in farming need not restrict diversity in the farmer's diet, and vice versa.

The discussion so far was not intended to imply that farming in all and every spatial setting necessarily patterns itself on the von Thünen's template of agribusiness. Nor was it intended to imply that agriculture everywhere and at all times was necessarily coupled with exchange. There are too many counterexamples – though none, as far as I am aware, is all that pretty. What I have tried to emphasize throughout this part of the discussion is that exchange, in one form or another, was a valid option, and at least some ancient agrarian communities took advantage of it early on. The relative abundance of ancient, however embryonic, cities dating to the earliest days off agriculture (some of which survive to this very day) bears testimony to that effect. Without taking advantage of division of labor and the capacity for exchange it is hard to imagine how a radical, almost complete, transition from hunting-gathering to agriculture (i.e., from eclectic acquisition to specialized production of food) could have been accomplished, of all species, by the most eclectic feeder of all.

Agriculture: summary

For a long time the ultimate environmental and behavioral causes for the sudden transition to agriculture remained poorly understood. The discussion in these last two chapters was essentially an attempt to fill the gap, if only in small measure. The inquiry was carried out in two steps. First, it was important to identify the most compelling inhibiting pressures that kept early people from making the same transition for more than 90% of the time that anatomically modern humans walked the earth. The discussion lists two such constraints: (1) risk and uncertainty across states of nature which is (and always was) the primary concern in any agrarian economy, and (2) the restriction on diet imposed by specialized food production which must have been, physiologically speaking, a serious concern to members of a species that evolved for millions of years in a highly eclectic feeding ecology.

With this understanding, the second logical step in the inquiry was to identify corresponding mitigating effects. In this respect, the discussion points to a coincidence of two distinct (though mutually reinforcing) sources traceable, respectively, to the physical environment and to human nature itself. On the level of the environment, it is clear now what was unclear and unexpected only a short decade ago: namely, that one of the a major features that distinguishes the climate of the last 10,000 years or so from earlier climates – either glacial or interglacial – is stability. Uncertainty across states of nature was thus largely mitigated. On the human level, nothing could be more handy at the onset of agriculture than a well-established propensity to exchange, for nothing could better reconcile the need for specialization in food production with the need for diversification in food consumption. Interestingly enough, the need for diversification – our inescapable (i.e., adaptive) dependence on overly eclectic diets – is probably a product of exchange to begin with. In this sense, ironically, exchange served to relax a constraint of its own making.

In sum, it is hard to see how the transition to agriculture could have been set in motion without a stable climate or, for that matter, without the human propensity to exchange. Climate stability and exchange are therefore two reasonable conditions for that transition. Standing alone, each provides only a necessary condition but, taken together, they seem to provide a sufficient one.

References

Aiello, L. C., & Wheeler, P. (1995). The Expensive-Tissue Hypothesis: The Brain and the Digestive System in Human and Primate Evolution. *Current Anthropology*, **36**, 199–221.

Alexander, R. D. (1979). *Darwinism and Human Affairs*. Seattle, WA: University of Washington Press.

Alexander, R. M. (1996). *Optima for Animals* (rev. ed.). Princeton: Princeton University Press.

Aschoff, J., Günther, B., & Kramer, K. (1971). *Energiehaushalt und Temperaturregulation*. Munich: Urban and Schwarzenberg.

Banning, E. B. & Köhler-Rollefson, I. (1992). Ethnographic Lessons for the Pastoral Past: Camp Locations and Material Remains near Beidha, Southern Jordan. In *Pastoralism in the Levant: Archaeological Materials in Anthropological Perspectives*, ed. O. Bar- Yosef & A. Khazanov, pp. 181–204. Madison: Prehistory Press.

Bar-Yosef, O. & Belfer-Cohen, A. (1989). The Origins of Sedentism and Farming Communities in the Levant. *Journal of World Prehistory*, **3**, 447–498.

Bar-Yosef, O. & Khazanov, A. (1992). Introduction. In *Pastoralism in the Levant: Archaeological Materials in Anthropological Perspectives*, ed. O. Bar-Yosef & A. Khazanov, pp. 1–10. Madison: Prehistory Press.

Bar-Yosef, O. (1994). The Contributions of Southwest Asia to the Study of the Origin of Modern Humans. In *Origins of Anatomically Modern Humans,* ed. M.H. Nitecki & D.V. Nitecki, pp. 23–66. New York: Plenum Press.

Bateman, A. J. (1948). Intra-sexual Selection in Drosophila. *Heredity*, **2**, 349–368.

Becker, G. S. (1976). *The Economic Approach to Human Behavior.* Chicago: The University of Chicago Press.

Becker, G. S. (1992). *Human Capital: A Theoretical and Empirical Analysis with Special Reference to Education.* Chicago: The University of Chicago Press.

Becker, G. S. (1993). Nobel Lecture: The Economic Way of Looking at Behavior. *Journal of Political Economy*, **101**, 385–409.

Belfer-Cohen, A. (1988). The Appearance of Symbolic Expression in the Upper Pleistocene of the Levant as Compared to Western Europe. In *L'Homme de Neandertal*, vol. v, *La Pens'ee*, ed. M. Otte, pp. 25–29. Liege: Etudes et Recherches Archeologiques de l'Universite de Liege.

Binford, L. R. (1968). Post-Pleistocene Adaptations. In *New Perspectives in Archaeology*, ed. S. R. Binford & L. R. Binford, pp. 313–341. Chicago: Aldine.

Binford, L. R. (1981). *Bones, Ancient Man and Modern Myths.* New York: Academic Press.

Binford, L. R. (1985). Human Ancestors: Changing Views of Their Behavior. *Journal of Anthropological Archeology*, **4**, 292–327.

Binford, L. R. (1992). Subsistence - a Key to the Past. In *The Cambridge Encyclopedia of Human Evolution*, ed. S. Jones, R. Martin, & D. Pilbeam, pp. 365–368. Cambridge: Cambridge University Press.

Binford, L.R. & Ho, C.K. (1985). Taphonomy at a Distance: Zhoukoudian, "The Cave Home of Beijing Man"? *Current Anthropology*, **26**, 413–442.

Bonner, J. T. (1965). *Size and Cycle: An Essay on The Structure of Biology*. Princeton: Princeton University Press.

Bourliere, F. & Hadley, M. (1970). The Ecology of Tropical Savannas. *Annual Review of Ecology and Systematics*, **1**, 125–152.

Bowlby, J. (1990). *Charles Darwin: A Biography*. London: Hutchinson.

Broca, P. (1878). Anatomie comparée des circonvolutions cérébrales. Le grand lobe Limbique et la scissure limbique dans la série des mammiféres. *Revue d'anthropologie*, **1**, 385–498.

Bryson, R. A. (1987). On Climates of the Holocene. In *Man and the Mid-Holocene Climatic Optimum*, Proceedings of the 17th Chacmool Conference, ed. N.A. Mckinnon & G.S.L. Stuart, pp. 1–13. Calgary, Alberta: the Archaeological Association of the University of Calgary.

Bunn, H. T. (1986). Patterns of Skeletal Representation and Hominid Subsistence Activities at Olduvai Gorge, Tanzania and Koobi Fora, Kenya. *Journal of Human Evolution*, **15**, 673–690.

Bunn, H. T. & Kroll, E. M. (1986). Systematic Butchery by Plio/pleistocene Hominids at Olduvai Gorge, Tanzania. *Current Anthropology*, **5**, 431–452.

Champion, T., Gamble, C., Shennan, S. & Whittle, A. (1984). *Prehistoric Europe*. London: Academic Press.

Childe, V.G. (1951a). *Man Makes Himself*. New York: New American Library.

Childe, V.G. (1951b). *Social Evolution*. New York: Henry Schuman.

Chivers, D.J. (1992). Diets and Guts. In *The Cambridge Encyclopedia of Human Evolution*, ed. S. Jones, R. Martin, & D. Pilbeam, pp. 60–64. Cambridge: Cambridge University Press.

Claiborne, R. (1970). *Climate, Man, and History*. New York: Norton.

Clark, C. W. (1976). *Mathematical Bioeconomics: The Optimal Management of Renewable Resources*. New York: John Wiley.

Clutton-Brock, T. H. (1977). Some Aspects of Intraspecific Variation in Feeding and Ranging Behaviour in Primates. In *Primate Ecology: Studies of Feeding and Ranging Behaviour in Lemurs, Monkeys and Apes*, ed. T. H. Clutton-Brock, pp. 539–556. London: Academic Press.

Coase, R. H. (1960). The Problem of Social Cost. *Journal of Law and Economics*, **3**, 1, 1–44.

Cohen, M. N. (1977). *The Food Crisis in Prehistory: Overpopulation and the Origins of Agriculture*. New Haven: Yale University Press.

Cronin, H. (1991). *The Ant and the Peacock: Altruism and Sexual Selection from Darwin to Today.* Cambridge: Cambridge University Press.

Crook, J. H. (1970). The Socio-Ecology of Primates. In *Social Behavior of Birds and Mammals*, ed. J. H. Crook, pp. 103–166. Academic Press.

Dansgaard, W., Johnsen, S. J., Clausen, H. B., Dahl-Jensen, D., Gundestrup, N. S., Hammer, C. U., Hvldberg, C. S., Steffensen, J. P., Sveinbjörnsdottirt, A. E., Jouzel, J. & Bond, G. (1993). Evidence for General Instability of Past Climate from a 250-kyr Ice-core Record. *Nature*, **364**, 218–220.

Darwin, C. (1868). *The Variation of Animals and Plants under Domestication.* Orange Judd.

Darwin, C. (1874). *The Descent of Man and Selection in Relation to Sex*, 2nd edn, Appleton.

Darwin, C. (1945). *Voyage Of The Beagle.* London: J. M. Dent & Sons.

Darwin, C. (1964). On the Origin of Species by Means of Natural Selection. Facsimile reprint of the first edition first published by J. Murry, 1859. Cambridge, MA: Harvard University Press.

Dawkins, R. (1987). *The Blind Watchmaker.* New York: Norton.

Dawkins, R. (1995). *River out of Eden.* New York: Basic Books.

Dawkins, R. (1996). *Climbing Mount Improbable.* New York: Norton.

Dawkins, R. & Krebs, J. R. (1979). Arms Races Within and Between Species. *Proceedings of the Royal Society of London*, **B 205**, 489–511.

Degabriele, R. (1980). The Physiology of the Koala. *Scientific American*, **243**, 94–99.

Derev'anko, A. P., ed. (1998). *The Paleolithic of Siberia: New Discoveries and Interpretations.* Chicago: University of Illinois Press.

Devereux, S. (1993). *Theories of Famine.* Brighton: Harvester Wheatsheaf.

DeVore, I. & Washburn, S. L. (1963). Baboon Ecology and Human Evolution. In *African Ecology And Human Evolution*, ed. F. C. Howell & F. Bourlière. pp. 335–367. Chicago: Aldine.

Diamond, J. (1997). *Guns, Germs, and Steel: The Fates of Human Societies.* New York: W. W. Norton.

Dott, R. H. Jr. & Batten, R. L. (1988). Evolution of the Earth, 4th edition. New York: McGraw-Hill.

Edozien, J. C. & Switzer, B. R. (1978). Influence of Diet on Growth in the Rat. *Journal of Nutrition*, **108**, 282–90.

Eiseley, L. (1961). *Darwin's Century.* Garden City, NY: Doubleday.

Estes, R. D. (1992). *The Behavior Guide to African Mammals.* Berkeley, CA: The University of California Press.

Ferris, S. D., Brown, W. M., Davidson, W. S. & Wilson, A. C. (1981). Extensive Polymorphism in the Mitochondrial DNA of Apes. *Proceedings of the National Academy of Sciences, USA*, **78**, 6319–6323.

Finley, M. I. (1973). *The Ancient Economy.* Berkeley: The University of California Press.

Fisher, R. A. (1958). *The Genetical Theory of Natural Selection.* New York: Dover.

FitzRoy, R. (1839). Narrative of the Surveying Voyages ... of the Globe. *Proceedings of the Second Expedition, 1831–1836, Under the Command of Captain Robert Fitz-Roy, R.N.,* vol. II. London: Henry Colburn.

Gamble, C. (1978). Resource Exploitation and the Spatial Patterning of Hunter-Gatherers: a Case Study. In *Social Organisation and Settlement,* ed. D. Green, C. Haselgrove & M. Spriggs, pp. 153–85. British Archaeological Reports S47, Oxford.

Gamble, C. & Soffer, O. (1990). Pleistocene Polyphony: The Diversity of Human Adaptations at the Last Glacial Maximum. In *The World at 18,000 B.P.,* vol. I, High Latitudes, ed. O. Softer & C. Gamble, pp. 1–23. London: Unwin Hyman.

Goldizen, A. W. (1990). A Comparative Perspective on the Evolution of Tamarin and Marmoset Mating Systems. *International Journal of Primatology,* **11,** 63–84.

Goodman, M. (1962). Immunochemistry of the Primates and Primate Evolution. *Annals of the New York Academy of Sciences,* **102,** 219–234.

Gordon, S. (1989). Darwin and Political Economy: The Connection Reconsidered. *Journal of the History of Biology,* **22,** Fall, 437–459.

Gould, S. J. (1993). Darwin and Paley Meet the Invisible Hand. In *Eight Little Piggies: Reflections in Natural History,* S. J. Gould. New York: W.W. Norton.

Hamilton, W. D. (1964). The Genetical Evolution of Social Behaviour, II. *Journal of Theoretical Biology,* **7,** 17–52.

Hamilton, W. D. (1967). Extraordinary sex Ratios. *Science,* **156,** 477–488.

Hardin, G. (1968). The Tragedy of the Commons. *Science,* **162,** 1243–1247.

Harding, R. S. O. & Strum, S. C. (1976). Predatory Baboons of Kekopey. *Natural History,* **85,** 46–53.

Harper, J. L. (1977). *Population Biology of Plants.* London: Academic Press.

Harris, D.R. (1992). Human Diet and Subsistence. In *The Cambridge Encyclopedia of Human Evolution,* ed. S. Jones, R. Martin, & D. Pilbeam, pp. 69–74. Cambridge: Cambridge University Press.

Hill, K. & Kaplan, H. (1994). On Why Male Foragers Hunts and Share Food. *Current Anthropology,* **34,** 701–706.

Hirshleifer, J. (1977). Economics from a Biological Viewpoint. *Journal of Law and Economics,* **XX,** 1, 1–52.

Hochman, O. & Ofek, H. (1977). The Value of Time in Consumption and Residential Location in an Urban Setting. *American Economic Review,* **67,** 5, 996–1003.

Hölldobler, B. & Wilson, E. O. (1994). *Journey to the Ants.* Cambridge, MA: Harvard University Press.

Houghton, J. (1997). *Global Warming: The Complete Briefing.* 2nd edn. Cambridge: Cambridge University Press.

Hu, F. S., Slawinski, D., Wright, H. E., Ito, E., Johnson, R. G., Kelts, K. R., McEwan, R. F., Boedighei, A. (1999). Abrupt Changes in North American Climate During Early Holocene Times. *Nature,* **400, 6743,** 437–440.

Immelmann, K. (1980). *Introduction to Ethology*. Plenum Press.

Immelmann, K. & Beer, C. (1992). *A Dictionary of Ethology*. Cambridge, MA: Harvard University Press.

Iregren, E. (1988). Finds of Brown Bear (*Ursus arctos*) in Southern Scandinavia – Indications of Local Hunting or Trade? In *Trade and Exchange in Prehistory: Studies in Honour of Berta Stjernquist*, Acta Archaeologica Lundensia, Series in 8°. N° 16, ed. B. Hårdh, L. Larsson, D. Olausson, & R. Petré, pp. 295–308. Stockholm: Almqvist & Wiksell.

Isaac, G. (1983). Aspects of human evolution. In *Essays on Evolution: a Darwin Century Volume*, ed. D. S. Bendall. Cambridge: Cambridge University Press.

Isaac, G. (1989). *The Archaeology of Human Origins*, ed. B. Isaac. Cambridge: Cambridge University Press.

Keynes, D. (ed.) (1988). *Charles Darwin's Beagle Diary*. Cambridge: Cambridge University Press.

Khazanov, A. M. (1984). *Nomads and the Outside World*. Cambrige: Cambridge University Press.

Klein, R. G. (1969). *Man and Culture in the Late Pleistocene: A Case Study*. San Francisco: Chandler.

Klein, R. G. (1989). *The Human Career: Human Biological and Cultural Origins*. Chicago: University of Chicago Press.

Klein, R. G. (1992). The Archeology of Modern Human Origins. *Evolution and Anthropology*, **1**, 5–14.

Kliks, M. (1978). Paleodietetics: A Review of the Role of Dietary Fiber in Preagricultural Human Diets. In *Topics in Dietary Fiber Research*, eds. G. A. Spiller and R. J. Amen, pp. 181–202. New York: Plenum Press.

Kummer, H. (1997). *In Quest of the Sacred Baboon: A Scientist's Journey*. Princeton, NJ: Princeton University Press.

Kurian, G. T. (1994). *Datapedia of the United States 1790–2000: America Year by Year*. Lanham, MD: Bernan Press.

Lambert, D. (1987). *The Field Guide to Early Man*. New York: Facts on File.

Leakey, M. D. (1975). Cultural Patterns in the Olduvai Sequence. In *After the Australopithecines*, eds. K. W. Butzer and G. Ll. Isaac, pp. 477–493. The Hague: Mouton.

Leakey, M. G., Spoor, F., Brown, F. H., Gathogo, P. N., Kiarie, C., Leakey, L. N. & Mcdougall, I. (2001). New Hominin Genus from Eastern Africa Shows Diverse Middle Pliocene Lineages. *Nature*, **410**, 433–440.

Lee, R. B. (1968). What Hunters Do for a Living; Or, How to Make Out on Scarce Resources. In *Man the Hunter*, ed. R. B. Lee & I. DeVore, pp. 30–48. Chicago: Aldine Press.

Lindly, J. M. & Clark, G. A. (1990). Symbolism and Modern Human Origins. *Current Anthropology*, **31**, 233–262.

Linton, R. (1956). *The Tree of Culture*. New York: Alfred A. Knopf.

Loenen, D. (1941). *Protagoras and the Greek Community*. Amsterdam: Noord-Hollandsche.

Lott, D. F. (1991). *Intraspecific Variation in the Social Systems of Wild Vertebrates*. Cambridge: Cambridge University Press.

Lovejoy, C.O. (1981). The Origin of Man. *Science*, **211**, 341–350.

Macdonald, D. W. (1984). *The Encyclopedia of Mammals*. New York: Facts on File.

MacLean, P. D. (1973). *A Triune Concept of the Brain and Behaviour*. Toronto: The University of Toronto Press.

Malthus, T. R. (1976). *An Essay on the Principle of Population*. New York: Norton.

Margulis, L. (1981). *Symbiosis in Cell Evolution*. New York: W.H. Freeman.

Marks, J. (1996). Molecular Anthropology in Retrospect and Prospect. In *Contemporary Issues in Human Evolution*, ed. Meikle, W. E., Howell, F. C. & Jablonski, N. G. San Francisco: California Academy of Sciences (Memoir 21), pp. 167–186.

Marshall, A. (1961). *Principles of Economics* (first published 1920). London: Macmillan.

Martin, K. & Voorhies, B. (1975). *Female of the Species*. New York: Columbia University Press.

Mauss, M. (1967). *The Gift: Forms and Functions of Exchange in Archaic Societies*. New York: Norton.

Maynard Smith, J. (1978). *The Evolution of Sex*. Cambridge: Cambridge University Press.

Mayr, E. (1963). *Animal Species and Evolution*. Cambridge, MA: Harvard University Press.

Milton, K. (1987). Primate Diets and Gut Morphology: Implications for Hominid Evolution. In *Food and Evolution: Toward a Theory of Human Food Habit*, ed. M. Harris & E. B. Boss, pp. 96–116. Philadelphia: Temple University Press.

Milton, K. (1995). Comment. *Current Anthropology*, **36**, 214–215.

Mincer, J. (1974). *Schooling, Experience, and Earnings*. New York: National Bureau of Economic Research, distributed by Columbia University Press.

Moehlman, P. D. (1989). Intraspecific Variation in Canid Social Systems. In *Carnivore Behavior, Ecology and Evolution*, ed. J. L. Gittleman, pp. 143–163. Ithaca: Cornell University Press.

Morgan, E. (1994). *The Scars of Evolution*. Oxford: Oxford University Press.

Napier, J. (1993). *Hands*. Princeton: Princeton University Press.

Nelson, R. A. (1975). Implications of Excessive Protein. In *Proceedings Western Hemisphere Nutrition Congress IV*, ed. P. L. White & N. Selvy, pp. 71–76. Acton, MA: Publishing Sciences Group.

Ofek, H. & Merrill, Y. (1997). Labor Mobility and the Formation of Gender Wage Gaps in Local Markets. *Economic Inquiry*, **25**, **1**, 28–47.

Osborn, H. F. (1929). *From the Greek to Darwin*. New York: Scribner's Sons.

Oster, G. F. & E. O. Wilson (1978). *Caste and Ecology in the Social Insects*. Princeton: Princeton University Press.

Papez, J .W. (1937). A Proposed Mechanism of Emotion. *Archives of Neurology and Psychiatry*, 38, 725–743.

Pellew, R. A. (1984). The Feeding Ecology of a Selective Browser, the Giraffe (*Giraffa camelopardalis*). *Journal of Zoology*, 202, 57–81.

Penfield, W. & Rasmussen, T. (1950). *The Cerebral Cortex of Man: A Clinical Study of Localization of Function*. New York: Macmillan.

Pfeiffer, J. E. (1982). *The Creative Explosion: An Inquiry into the Origins of Art and Religion*. New York: Harper & Row.

Pianka, E. R. (1970). On r- and K-selection. *American Naturalist*, **104(940)**, 592–597.

Plato (1978). Protagoras. In *The Collected Dialogues of Plato*, ed. E. Hamilton and H. Cairns, pp. 308–352. Princeton: Princeton University Press.

Plutarch (1979). *The Lives of the Noble Grecians and Romans*, ed. J. Dryden & A. H. Clough. New York: Random House.

Potts, R. B. (1988). *Early Hominid Activities at Olduvai*. Chicago: Aldine.

Potts, R. B. & Shipman, P. (1981). Cutmarks Made by Stone Tools on Bones from Olduvai Gorge, Tanzania. *Nature*, 291, 577–580.

Raynaud, D., Jouzel, J., Barnola, J. M., Chappellaz, J., Delmas, R. J. & Lorius, C. (1993). The Ice Record of Greenhouse Gases. *Science*, 259, 926–934.

Redford, R. A. (1945). The Economic Organization of a POW Camp. *Economica*, November, 189–201.

Reichard, U. (1995). Extra-pair Copulations in Monogamous Wild White-handed Gibbons (*Hylobates lar*). *Zeitschrift für Säugetierkunde*, 60, 186–188.

Rindos, D. (1984). *The Origins of Agriculture: An Evolutionary Perspective*. New York: Academic Press.

Robin, E. D. (ed.) (1979). *Claude Bernard and the Internal Environment*. New York: Marcel Dekker.

Rosenberg, N. & L. E. Birdzell, Jr. (1986). *How the West Grew Rich: the Economic Transformation of the Industrial World*. New York: Basic Books.

Schabas, M. (1994). The Greyhound and the Mastiff: Darwinian Themes in Mill and Marshall. In *Natural Images in Economic Thought*, ed. P. Mirowski. Cambridge: Cambridge University Press.

Schiappa, E. (1991). *Protagoras and Logos: A Study in Greek Philosophy and Rhetoric*. University of South Carolina Press.

Schidlowski, M. (1988). A 3,800–million-year Isotopic Record of Life from Carbon Sedimentary Rocks. *Nature*, 333, 313–318.

Schild, R. (1984). Terminal Paleolithic of the North European Plain: A Review of Lost Chances, Potential and Hopes. *Advances in World Archaeology*, 3, 193–274.

Schweber, S. S. (1978). The Genesis of Natural Selection – 1838: Some Further Insights. *BioScience*, 28, 321–326.

Schweber, S. S. (1980). Darwin and the Political Economists: Divergence of Character. *Journal of the History of Biology,* **13**, 195–289.

Sen, A. (1981). *Poverty and Famines.* Oxford: Clarendon Press.

Sheldon, J. W. (1992). *Wild Dogs: The Natural History of the Nondomestic Canidae.* San Diego, CA: Academic Press.

Shipman, P. (1986). Scavenging or Hunting in Early Hominids. *American Anthropologist,* **88**, 27–43.

Sibley, C.G. (1992). DNA-DNA Hybridization in the Study of Primate Evolution. In *The Cambridge Encyclopedia of Human Evolution,* ed. S. Jones *et al.*, pp. 313–315. Cambridge: Cambridge University Press.

Smith, A. (1976). *The Wealth of Nations.* Chicago IL: The University of Chicago Press.

Smith, B. D. (1995). *The Emergence of Agriculture.* New York: Scientific American Library.

Soffer, O. (1994). Ancestral Lifeways in Eurasia - the Middle and Upper Paleolithic Records. In *Origins of Anatomically Modern Humans,* ed. M.H. Nitecki & D.V. Nitecki, pp.101–119. New York: Plenum Press.

Strathern, M. (1992). Qualified Value: The Perspective of Gift Exchange. In *Barter, Exchange and Value: an Anthropological Approach,* ed. C. Humphrey *et al..* Cambridge: Cambridge University Press.

Stringer, C. & Gamble, C. (1993). *In Search of the Neanderthals: Solving The Puzzle of Human Origins.* New York: Thames and Hudson.

Szabó–Patay, J. (1928). A Kapus-hangya. *Természettudományi Közlöny,* Budapest, 1928: 215–219.

Trivers, R. (1971). The Evolution of Reciprocal Altruism. *Quarterly Review of Biology,* **46**, 35–57.

Trivers, R. (1972). Parental Investment and Sexual Selection. In *Sexual Selection and the Descent of Man 1871–1971,* ed. B. Campbell, 136–179. Chicago: Aldine.

Trivers, R. (1985). *Social Evolution.* Menlo Park, CA: Benjamin/Cummings.

Tullock, G. (1994). *The Economics of Non-Human Societies.* Tucson, AR: Pallas Press.

Vander Wall, S.B. (1990). *Food Hoarding in Animals.* Chicago: University of Chicago Press.

von Thünen, J. H. (1966). *Isolated State,* an English translation of *Der isolierte staat* by C. M. Wartenberg, ed. P. Hall. Oxford: Pergamon Press (originally published in 1826).

Wallace, A. R. (1869). Geological Climates and the Origin of Species. *Quarterly Review,* **126**.

Wallace, A. R. (1870). *Contributions to the Theory of Natural Selection.* London: Macmillan.

Wallace, A. R. (1889). *Darwinism: An Exposition of the Theory of Natural Selection.* London: Macmillan.

Wallace, A. R. (1908). *My Life: A Record of Events and Ideas.* London: Chapman & Hall.

Weeks, P. (1999). Interactions Between Red-billed Oxpeckers, *Buphagus erythrorhynchus,* and Domestic Cattle, *Bos taurus,* in Zimbabwe. *Animal Behviour,* **58**, 1253-1259.

Weiner, S., Xu, Q., Goldberg, P., Liu, J. & Bar-Yosef, O. (1998). Evidence for the Use of Fire at Zhoukoudian, China. *Science*, **281**, 251–253.

Wilson, D. S. (1992). Group Selection. In *Keywords in Evolutionary Biology*, ed. E. F. Keller & E. A. Lloyd. Cambridge, MA: Harvard University Press.

Wilson, D. S. & Sober, E. (1994). Reintroducing Group Selection to the Human Behavioral Sciences. *Behavioral and Brain Sciences*, **17**, 587–654.

Wilson, D. S. & Sober, E. (1998). *Unto Others: The Evolution and Psychology of Unselfish Behavior*. Cambridge, MA: Harvard University Press.

Wilson, E. O. (1975). *Sociobiology: The New Synthesis*. Cambridge, MA: Harvard University Press.

Wilson, E. O. (1992). *The Diversity of Life*. New York: Norton.

Wilson, E. O. (1998). *Consilience: The Unity of Knowledge*. New York: Alfred A. Knopf.

Wobst, H. M. (1990). Minitime and Megaspace in the Palaeolithic at 18k and Otherwise. In *The World at 18,000 B.P.*, vol. II, *Low Latitudes*, ed. O. Softer & C. Gamble, pp. 322–334. London: Unwin Hyman.

Wolfe, D. W. & Erickson, J. D. (1993). Carbon Dioxide Effects on Plants: Uncertainties and Implications for Modeling Crop Response to Climate Change. In *Agricultural Dimensions of Global Climate Change*, ed. H. M. Kaiser & T. E. Drennen, pp. 153–178. Delray Beach, FL: St. Lucie Press.

Write, C. (1870). Review of Wallace's "Contributions to the Theory of Natural Selection." *North American Review*, **111**, 282–311.

Wynne-Edwards, V. C. (1963). Intergroup selection in the evolution of social systems. *Nature* 200: 623–626.

Zihlman, A. L. (1983). A Behavioral Reconstruction of Australopithecus. In *Hominid Origins: Inquiries Past and Present*, ed. Reichs, K. J. pp. 207–238. Washington, DC: University Press of America.

Index